THE COMPREHENSIVE GUIDE
TO BOARD WARGAMING

To My Family

THE COMPREHENSIVE GUIDE TO
Board Wargaming

Nicholas Palmer

HIPPOCRENE
BOOKS, INC.

NEW YORK, N.Y.

Hippocrene Books, Inc.
171 Madison Avenue
New York, N.Y. 10016

Library of Congress Cataloging in Publication Data

Palmer, Nicholas.
The comprehensive guide to board wargaming.
1. War games. I. Title.
U310.P35 1977 793'.9 76–53218
ISBN 0–88254–430–6

Printed in England

CONTENTS

CONTENTS CONTINUED

ACKNOWLEDGEMENTS

Whenever one opens a book, be it *Little-Known Hypotheses on Diffuse Topologies* or *Lust-Crazed Murderers I Have Known*, there always seems to be an introductory note saying that without the help of dozens of associates this literary masterpiece would not have been possible. The nature of the help is rarely specified, and the general impression conveyed is that the author is indulging in becoming modesty.

This is not the case here: the book would not have appeared, or only in truncated form, without the assistance and support of a number of people. The idea started over a *Diplomacy* game. A fiend named Simon Dally had just ruined my excellent position for no better reason than that it enabled him to win. In the course of the post-mortem, it transpired that he was a publisher. I suggested in an idle fancy that it was time a book was written on board wargames. He invited me to submit a synopsis for consideration, and the idea was borne through to fruition, made possible by the ex-fiend's help, advice, and encouragement at every turn.

James Dunnigan and Richard Berg of Simulations Publications Incorporated helped me in every way, with numerous sample games, much helpful advice, and an absolute disinclination to ask for special consideration for SPI games; they in fact suggested the exclusion of an SPI game mentioned in the synopsis, feeling it was not one of their best.

Thomas Shaw of Avalon Hill, Marc Miller of Game Designers' Workshop and Lou Zocchi of Gamescience Inc. also gave me every assistance, and the Attack Wargaming Association gave me advance notice of future games. Graeme Levin and *Games and Puzzles*, for whom I have the pleasure of working as wargames editor, gave me invaluable help in obtaining games.

Charles Vasey and Geoff Barnard, co-editors of *Perfidious Albion*, contributed respectively numerous game reviews and the rules for 'Situation 21 *Panzerleader*'. Without Charles Vasey's help, the coverage of the wide range of wargames in print would have been much less comprehensive. Marcus Watney and Mark Gleeson arranged the photographs which were patiently taken by Geoff Goode.

Finally, Christian Strachan helped me through numerous playtests – and usually won, but we won't dwell on that.

My grateful thanks to all of them.

References to rules, illustrations and examples of play made with permission granted by:

The Avalon Hill Game Company, 4517 Harford Road, Baltimore, Maryland 21214, USA.

Game Designers' Workshop, 203 North Street, Normal, Illinois 61761, USA.

Philmar Ltd, 47–53 Dace Road, London E3 2NG, England.

Simulations Publications Incorporated, 44 East 23rd Street, New York, NY 10010, USA.

INTRODUCTION

It takes some time before writing catches up with new hobbies, and books are published on the subject; the enthusiasts are busy exploring their new domain, and nobody else knows much about it. Nevertheless, board wargaming is now over twenty years old in its present form and both players and games continue to multiply, with over three hundred designs currently on the market and more appearing every month; it seems time that a guide to the field be made available.

This book is written both for experienced players and newcomers to wargames. Some compromises have had to be made, and I hope that the former will excuse the inclusion of points which seem obvious to them. The first two chapters are mainly intended for readers not familiar with the games and their history. The book then divides into two sections. Parts I–III are designed to demonstrate all the main aspects of good play, from the planning of general strategy down to the nitty-gritty of tactical detail. Examples in each chapter are from different games, so that the reader gradually becomes acquainted with a wide range of simulations of different types, from grand strategy to tactical manœuvre, in different eras, and on land, sea and air. The illustrations are mostly from actual play, to give the best impression of the games in practice. Finally, there are six problems for the reader to challenge his understanding of the concepts in each chapter in Parts I and II.

Part IV, a third of the total text, is a guide to virtually every wargame generally available at the time of publication. The major companies have given invaluable help in predicting their new games for the first half of 1977, while the book is being printed, so that the tour of the wargames in print could be as up to date as possible. A short, and I hope unbiased, review is given of all the games on which I have detailed information. Coupled with opinion polls from two leading hobby magazines, the reviews should make it possible for the reader to find the subject, scale and type of wargame which suits him best; buying games from their titles and manufacturer's descriptions alone can be a disappointing and expensive way to find what one really wants.

Part V shows a sample game, blow by blow.

Wargames, like any hobby, should first and foremost be fun. I have no doubt that some of the ideas and comments in this book will be disputed by other experienced players with different playing styles. But I hope that they will none of them be bored, for I have tried to infuse the book with the excitement and the absorbing interest of wargames, and with such a rewarding subject to write about it is hard to go altogether wrong.

Nicholas Palmer
c/o Arthur Barker Ltd,
11 St John's Hill,
London SW11 1XA

Preliminaries

1 THE WARGAMES EXPLOSION

There was a time, just a generation ago, when wargames were unknown to most people, except perhaps as a semi-secret planning device used by defence planners in government. Yet by 1975 sales reached an annual rate of nearly three-quarters of a *million* games, a 600% increase over just five years ago. There are now thought to be well over a hundred thousand active players, and each month brings in a further flood of new players and games. In this chapter, we shall try not only to describe the many kinds of wargame, but also to explain this astonishing explosion of interest.

Board wargames are first and foremost games: they are designed for entertainment, and therefore differ from their military equivalents. At the same time, every effort is made to parallel real life sufficiently closely to give the players the feeling that they are experiencing the same sort of problems which were faced by Napoleon, Patton, Rommel, Zhukov, Nelson and the other past, present and even future commanders whose battles are simulated. Chess is a kind of wargame, designed with a vague basis in early warfare, but it is stylized and abstract. A board wargame usually simulates a particular battle or campaign, and tries to incorporate every aspect of it, up to the point where the simulation would become so complicated as to cease being fun to play.

There are board wargames on nearly every conceivable type of conflict: on every major Second World War battle, Alexander's campaigns, the great naval battles, hypothetical Nato–Warsaw Pact confrontations, and science fiction campaigns set thousands of years in the future. The games may cover the whole world over a period of years, or they may focus on some microscopic clash in a tiny encounter with minute-by-minute action. Yet the basic principles of most wargames are the same, however different they are to play, and a player who enjoys one has a feast of varied delights in store. The main problem is that there are not enough hours in the day to play as many as one would like!

There are two kinds of wargame: the board game and the miniatures game. In this book, we shall be considering the former, but let us briefly examine the difference. A board game has a printed map with a grid on it to facilitate movement and direction of fire: instead of saying 'advance three miles south-east and engage the defenders in the village at the road intersection' one can just say 'move to grid reference V19 and attack V20'. The effect is a cross between a road map and a very large chess board. The units are represented by cardboard printed counters, and there are detailed rules for playing out the particular battle being simulated. The vast majority of commercially produced wargames are of the board type.

Miniature games, however, have a long tradition, and exist in countless versions all over the world, as well as providing the inspiration for some board games. The impetus for them has always been the desire of collectors of military models to have them *do* something, and a miniatures game basically consists of a set of rules for the use of military models in a game. The rules are fairly generalized and leave a number of questions to be decided by the players. While each type has its adherents who dislike the other, most players enjoy both. The advantage of miniature wargaming is that it can be highly attractive visually, with subtly camouflaged model tanks nosing through curving valleys on a sandtable; for those who like both modelling and games it provides an ideal meeting point. Board wargames have more in common with other board games like chess. Their great virtue is the tremendous range of subjects which can be covered. A miniatures game covering the whole Soviet–German front in the Second World War would involve a mind-boggling array of pieces and impossible complexity. For a board game there is nothing easier; the scale of the map and size of units represented by the counters is adjusted accordingly, with new rules to fit the context. Another attractive feature is the real challenge to skill, enhanced by detailed rules, which is comparable to chess or Go, without the need to memorize lengthy opening gambits. Skill in wargames is a subtle and semi-intuitive matter, because the same situation is unlikely to occur twice, and the outcome of individual battles can seldom be predicted with absolute certainty. By the time you have read this book, you should be able to take on average players with reasonable optimism about the outcome. But real expertise in any particular game comes only with practice, when you can glance at a position and say, 'Aha! *There* is the weak point!' –

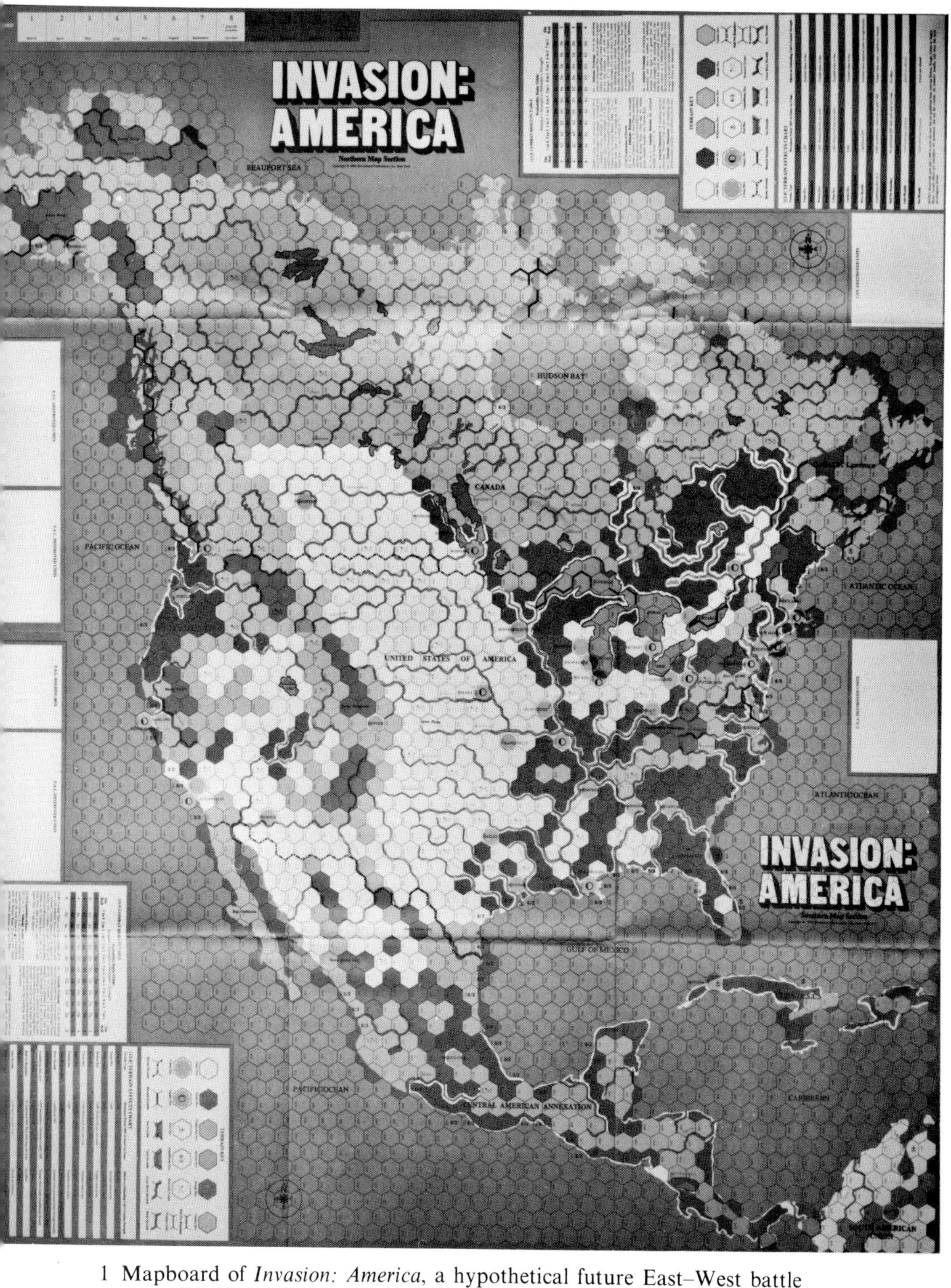

1 Mapboard of *Invasion: America*, a hypothetical future East–West battle on American soil.

not because you will have seen the position before (you won't) but because you will have developed a feeling for the essence of positions in that game.

In the rest of the book, we will use 'wargame' to refer to the board type, unless otherwise stated.

HISTORY OF WARGAMES

It is not really possible to pluck out of the past one particular game as the first wargame, although chess and Go must both be candidates for spiritual ancestry. In the medieval period, a great many experiments were made with the pieces and board of chess, in an attempt to make the game more relevant to contemporary warfare or, conversely, to use new ideas in military affairs to add interest to the game. It is said that the favourite game of the great Mongol Emperor, Timur (1366–1405), was 'Complete Chess', a variant incorporating such additional pieces as a Wazir, two Dabbabbas, two Scouts, two Camels and two Giraffes. Somewhat more relevant to wargames was a game invented in 1780 by Helwig, Master of the Pages of the Duke of Brunswick. Helwig's job seems to have left him with a fair amount of free time, as he settled on a game with 1666 squares and 120 fighting pieces, including infantry, cavalry and artillery (each with different movement rules). Specialized rules for pontoon and entrenchment construction appeared. Helwig's game resembled real contemporary warfare much more than chess, but the basis was still an abstract conflict in an imaginary area. This was developed by Georg Vinturinus, a military writer from Schleswig, into a more complex game on an actual map, covering the Franco-Belgian border. This board had 3600 squares, and a number of innovations, including military supplies and communication lines. This was in 1798, and the resemblance to current board wargames is marked. Had the technical means of mass production been available at that time, it is just conceivable that the idea would have taken hold. As it was, such games continued to be regarded as interesting personal variations on chess. The impetus of wargame development fell to the Prussian army, and in particular a Lieutenant von Reisswitz.

Von Reisswitz's father had developed a game which strongly resembled miniature wargames: it was played on a sandtable, with a particular scale, 1:2373, and the use of rulers to work out distances rather than a square grid system. This game was developed by von Reisswitz junior in the 1820s with realistic military-type maps and detailed rules. He succeeded in impressing General von Muffling, then Chief of Staff, with his brainchild at a demonstration in 1824. 'It is not a game at all!' von Muffling is said to have exclaimed. 'It is a training for war!' Von Muffling arranged for every regiment in the army to be furnished with a set, and urged them to practise with it.

The game was gradually developed over the ensuing half-century, with a steady trend *away* from the board wargame hallmark of detailed and precise rules, and increasing responsibility being given to an umpire. In 1876, this movement was accelerated by a new version of the game developed by Colonel von Verdy du Vernois. His game leant extremely heavily on the umpire and could be said to have become a framework for the players and the umpire to test their wits and expertise on each other. This version was called 'free *Kriegspiel*', as distinct from the traditional, though still fairly free by our standards, 'rigid *Kriegspiel*'.

After the Prussian triumph against the French in 1870, the armies of the world keenly studied the Prussian methods to try to find the formula for victory. Among the distinctive features was *Kriegspiel*, and all the major countries experimented with it in the years leading up to the First World War. The games were credited with a contribution towards a number of successes, notably the Japanese victory over Russia in 1904–5, for which the Japanese had carefully prepared with wargames. That the system had its limitations was shown by the German preparations for the First World War. The Chief of the General Staff up to 1906, Graf von Schlieffen, developed the basic plan which would be used in the initial assault, and tested it thoroughly with wargames. However, all the assumptions which he made in the plan were also made in the game, and the games dutifully confirmed that the plan was splendid, a process known to computer programmers as 'garbage in, garbage out'. This shows how *not* to design a wargame: to be a useful tool for simulation rather than mere entertainment, the game should only incorporate those facts which are definitely known, with possible variations tested for by playing the game with different assumptions. Most board wargames now have a variety of scenarios, even in the historical games, known as 'what-ifs', e.g. 'what if Britain had not intervened?' Of course, hindsight is easy. Von Schlieffen probably thought that such beliefs as that Britain would not intervene *were* soundly based on fact.

In the Second World War the Germans were again the main users of wargames, and in at least three cases they obtained important results. In 1939, they showed that it was possible to make a speedy breakthrough in the Ardennes to turn the Maginot Line. After the fall of France, a game showed numerous probably insuperable obstacles to a successful invasion of Britain, a judgement generally confirmed by present-day historians. When the Soviet Union was attacked, the operation was again wargamed in detail in advance.

One German wargame, in November 1944, had the unique feature that it became real life. The 5th Panzer Army defending Germany's western approaches were in the process of wargaming an anticipated American attack when it actually started to happen. Model, the Army Group commander in the area, ordered that the game be continued with up-to-the-minute news from the front, and the decisions resulting from the game were rushed off to the front for application.

Since the war, the United States has been the pace-maker in wargames, and the advent of computer technology and sophisticated communications equipment has brought about a swing back towards 'rigid' wargaming, with well-defined rules and detailed calculation systems replacing the freewheeling and somewhat arbitrary umpire systems. The post-war period also saw the reintroduction of board wargames for entertainment.

The first of these was produced by Charles Roberts in 1953: *Tactics*. This was an abstract game of contemporary military combat (its successor, *Tactics II*, is still on sale), and could be said to take up the thread left by Vinturinus and Helwig over 150 years earlier. Despite amateur production and distribution, Roberts succeeded in selling 2000 copies, and in 1958 he decided to take the plunge into professional game design. He set up the Avalon Hill Company, which to this day remains one of the two giants of the wargaming field, although it is now under different management. The first 'theme' (i.e. dealing with a particular battle) wargame published by Avalon Hill was *Gettysburg*. Although the American Civil War has not turned out quite as popular a subject for simulation as modern warfare (probably because there were fewer variables in the less complex nineteenth-century conflict, so that modern themes give a more intricate game), *Gettysburg* was a success and is still played, nearly twenty years and umpteen innovations later. The next few years saw the emergence of two more of what would become known as the 'classic' line of wargames: *Stalingrad* and *Waterloo*. The 'classic' games are fairly simple and fast-moving, with similar rules of play, but each succeeds (mainly by varying unit speeds and terrain) in giving a distinct flavour of the period. *Stalingrad* is especially successful in providing the 'feel' of the strategic problems on the German–Soviet front, despite a considerable disregard for fine points of historical detail. Expert players tend to regard the continuing survival of the 'classics', in the face of far more sophisticated newcomers, with the irritation of literary critics who see Harold Robbins outselling Tolstoy; however, although one would miss a lot by playing only 'classics', they do retain freshness and excitement even after years of play.

It was not until the end of the sixties that Avalon Hill (generally referred to as AH) encountered serious competition. The moving spirit was a former AH designer, James Dunnigan. His company, Simulations Publications Incorporated (SPI), sold their first games in 1969, at a time when AH's sales were near 100,000. By 1972, SPI was selling at the rate of 150,000 games a year, although AH were still ahead. In 1975, SPI estimated that their sales had reached 385,000, to just under 300,000 by AH.

One of the secrets of SPI's success is a magazine, *Strategy and Tactics*, which appears bimonthly with a new game in every issue; these games are later sold individually as well, as are other games. The total number of SPI wargames is correspondingly enormous: there are currently over seventy-five in print, compared with about twenty-five from Avalon Hill, who conversely usually have higher print runs of each game.

As far as standards go, it is difficult to be dogmatic (not that that stops people) when in some cases the same people have designed games at different times for each company. SPI's vast output does not seem to have negative effects on thoroughness in playtesting and rule formulation; nor does AH's comparatively narrow range appear to inhibit it from experiments and innovations. Nevertheless, there is a distinct difference in style between the companies. Many SPI games have been shorter than the average AH product and playable in a few hours compared with a typical 4–6 hours for AH games (recent designs by both companies have tended to eliminate the difference by providing scenarios ranging in length from a couple of hours to – almost! – infinity). SPI's games are noted for their historical detail and frequently varied unit types; they are also substantially cheaper as a rule. AH's products have better physical quality: the board is mounted, the boxes are mostly in a convenient bookcase format, and the general effect is often highly colourful, with even the rules in an attractive and handy booklet. There has been a recent trend for AH to buy successful game designs from small companies, revamp them in a new edition, and use the company sales network to bring them to a wider audience.

Critics of SPI have claimed that the company is no longer refreshing their games with as many new ideas as in their early days; others argue that AH concentrates too much on packaging, resulting in unnecessarily high prices. Each defends itself vigorously, and in the author's view it is in fact impossible to say whose games are the better buy: the balance of advantage is a matter of taste, and the vast majority of players have a number of favourite games from both companies, and others.

In the last few years, in particular, a number of other wargame manufacturers have appeared, the most widely known being Battleline, who produce carefully-researched games, often at the tactical end of the scale, and Game Designers' Workshop, who specialize in the increasingly popular 'monster' game type with over 1000 counters. They are above all famous for their gigantic Second World War game, *Drang Nach Osten.* This *divisional* level simulation of the entire Soviet front is immensely admired by 'hard-core' players for its fascinating breadth and complexity. Like its rival on the same subject, SPI's *War in the East*, it is hard to finish unless you settle down on a desert island with an equally dedicated opponent and no tiresome distractions like work, sleep or social life.

I am indebted to SPI for permission to draw heavily on 'The History of Wargaming', by Martin Campion and Steven Patrick in *Strategy and Tactics 33*, for much of the historical account in this section.

CURRENT TRENDS

Two tendencies are clearly visible in wargame design. One is to very big games, with short scenarios provided for players with less time. A distinction should be (but rarely is) made here between games which are big because of all the variables (economics, politics, production, varied capabilities, supply, etc.) involved, and those which involve smaller units than might be expected, e.g. divisions on a whole front, like *Drang Nach Osten.* The first type is the more unusual: it has the advantage of presenting an exceptional challenge to the skill of both players with a simulation as near reality as possible, minus the bits one can do without, such as digging trenches and getting shot. The second type is easier to play in practice (though very likely lengthier), and has the virtue of mixing strategy and tactics in a single game, so that one not only determines the plan for the capture of the whole enemy country, but also supervises the individual battles.

The other trend is towards political rules in wargaming, especially with multi-player games. The forerunner here has developed a cult following: *Diplomacy*, now distributed in the USA by AH. This seven-player game has a highly abstract combat system, and is unimpressive as a military simulation. Its special attraction lies in the cross-currents between players, with an explicit invitation in the rules to conduct negotiations in the grand tradition of Machiavelli.

The problem in political multi-player games is: what is to be done about history? Should a Franco-German alliance in the Second World

War, for instance, be allowed? One school of thought, which dominated in the design of e.g. *Third Reich* and *World War III*, holds that alliances in historical or present-day games must be reasonably plausible, as otherwise the players will lose the link with real life which gives wargames part of their interest. Even if one disagrees with this, one does not necessarily have to go to the 'history is bunk' extreme in the opposite direction, as in, e.g., *Strategy I*, which attempts to extract the military essence from wars through the ages, without much regard for political constraints. There will always have been reasons (whether logical or not) for history to have turned out as it did: for instance, a Franco-German alliance in the Second World War was unlikely (apart from governmental hostility), basically because French public opinion would almost certainly not have stood for it. In my view, these constraints and no others should be reflected in the game. Thus, in the example, it would be permissible for the French player to ally with Germany, but there would be a high risk of civil disorder immediately paralysing the country. The inducement, therefore, would have to be something magnificent, and it is hard to think of any arrangement which would make it a good idea for the French player. But the possibility would be there, and circumstances would sometimes arise when the gamble might seem worthwhile.

It was a postal game of *Diplomacy*, incidentally, which led to the most alarming transition to real life in recent wargames history. A Manchester player was France, and had conducted lengthy negotiations with Germany about tactics against England, concluding with a telegram from his ally: 'ATTACK ON LIVERPOOL AGREED'. This was at the height of an IRA bombing campaign in England, and the unfortunate 'Frenchman' was questioned by the police in exhaustive detail before they dubiously accepted his story.

The alternative to political rules is set political conditions, and many games have a number of alternative scenarios representing different backgrounds to the conflict. *Sinai*, a simulation of the Arab–Israeli wars, has a set of scenarios for each conflict, representing different political conditions: increased Arab cooperation, slow Israeli reaction, etc. (there is even an 'Arab fantasy scenario' in 1967 in which everything goes right for them; the designers think Israel would *still* have won). The advantage of this approach is that it enables players to get on with their game without worrying about losing through some devious diplomatic manœuvre. In *Diplomacy*, it is impossible to win against competent opposition without political success in gaining allies. In more typical wargames, you can and will win by 'military' skill.

TYPICAL WARGAMERS

There are few more depressing experiences than being told that every member of a group to which one belongs is 'really the same', whether the generalization is supposed to apply to all Frenchmen, all Latter-Day Adventists, or all bridge-players. The most important thing about the typical wargamer is that he does not exist: wargamers come in all shapes and sizes. Nevertheless, it is possible to identify some characteristics common to many wargamers; because this helps to show the nature of the attraction of wargames it is worth trying to do so.

There is a fairly strong resemblance to chess players. Wargamers also have to be willing to spend a whole evening staring at a board with their brains whirring, calculating the ideal strategy for the frustration of the fiendish schemes of the fellow opposite. Accordingly, wargamers tend to be quiet and slightly introverted (with some very definite exceptions); it may be possible to play *Drang Nach Osten* while singing 'Roll Out the Barrel' and downing a case of beer, but it is rare. The great majority of players are male, as in chess, and there is a large proportion of students and other intellectually oriented types, though this is probably less marked than in chess, perhaps because chess is more abstract.

Just as wargamers vary, so do their reasons for enjoying the hobby, but most players would probably agree that there are three basic attractions compared with other games like chess, Go and 'family games'. There is a high skill level requiring little knowledge in the way of memorized openings. There is the challenge in the historical games of changing history, out-generalling the great commanders from the depths of one's armchair. And there is the tension and excitement arising from the uncertainty which is always present: one knows the range of results which individual battles in a game can have, and one can be fairly sure of the general trend of results in a general offensive, but each battle may have several possible results, and the best player is always nagged by the feeling that he has not allowed for every possibility.

2 BASICS OF WARGAMING

THE BOARD

The most basic feature of a board wargame is, of course, the board. We have seen a picture of one of these (for *Invasion: America*) in the preceding chapter, and all boards have a number of common features, though some are much simpler.

Every board consists of a map, on which a grid of six-sided spaces (hexagons) is superimposed, with occasional exceptions where the grid uses large areas instead for more strategic movement. This basic similarity is modified by the enormous range of maps. There are maps which are entirely blank, representing flat terrain, a naval or aerial game, or possibly a lazy designer. There are maps entirely covered with every conceivable type of terrain feature, from towering mountains to thick jungle. There are maps covering the entire galaxy, and maps describing a village in minute detail. Which map is chosen is the decisive factor for the *scale* of the game and also, less obviously, for how it will 'feel' to play. For instance, if you open a game box and see a map like the (much-admired) one used in *Anzio*, which is dominated by the green and brown of forest and rough terrain, then you know at once that this is a game featuring tough fighting for key positions in the terrain (roads through the hills, dominating positions over beaches, and defensive strongpoints in mountains). If it looks more like the open country dominating the *Rocroi* map, then the game is likely to feature fast movement and broad positional sweeps, with the emphasis on tactical manœuvres. In general, the games with most terrain will be the 'operational' (regiment/division) level ones, where it plays a major part in planning; at a tactical level, all the units will very likely be in the same sort of terrain, while strategic games often have a time-scale of months per move, making obstacles like medium-sized rivers pretty insignificant.

Newcomers to wargaming are sometimes taken aback by the hexa-

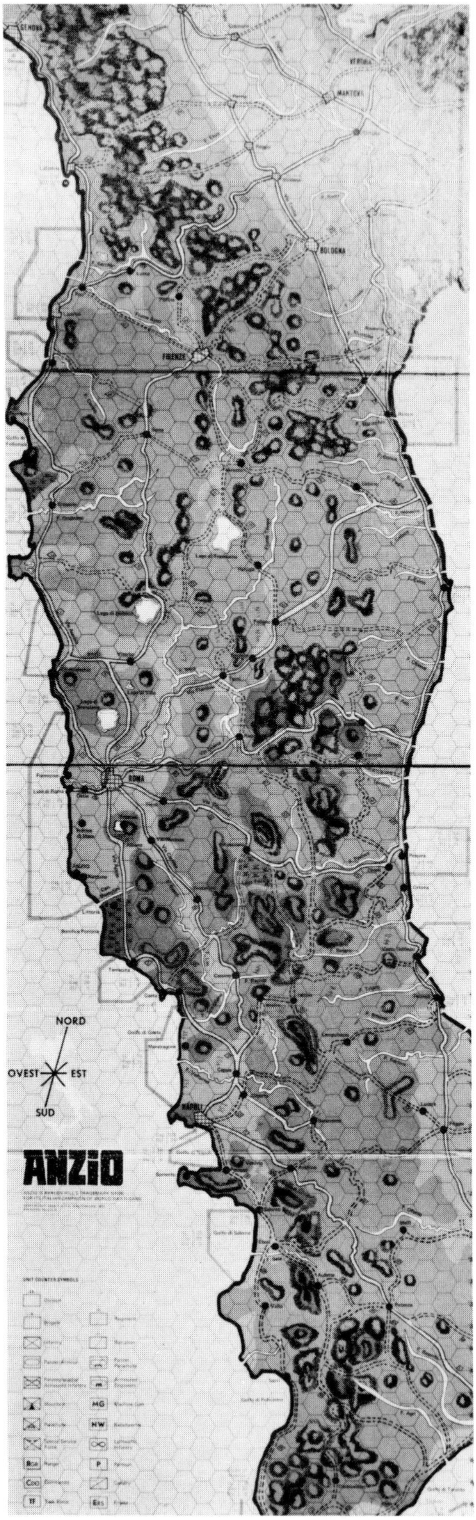

2 The mapboard of *Anzio*.

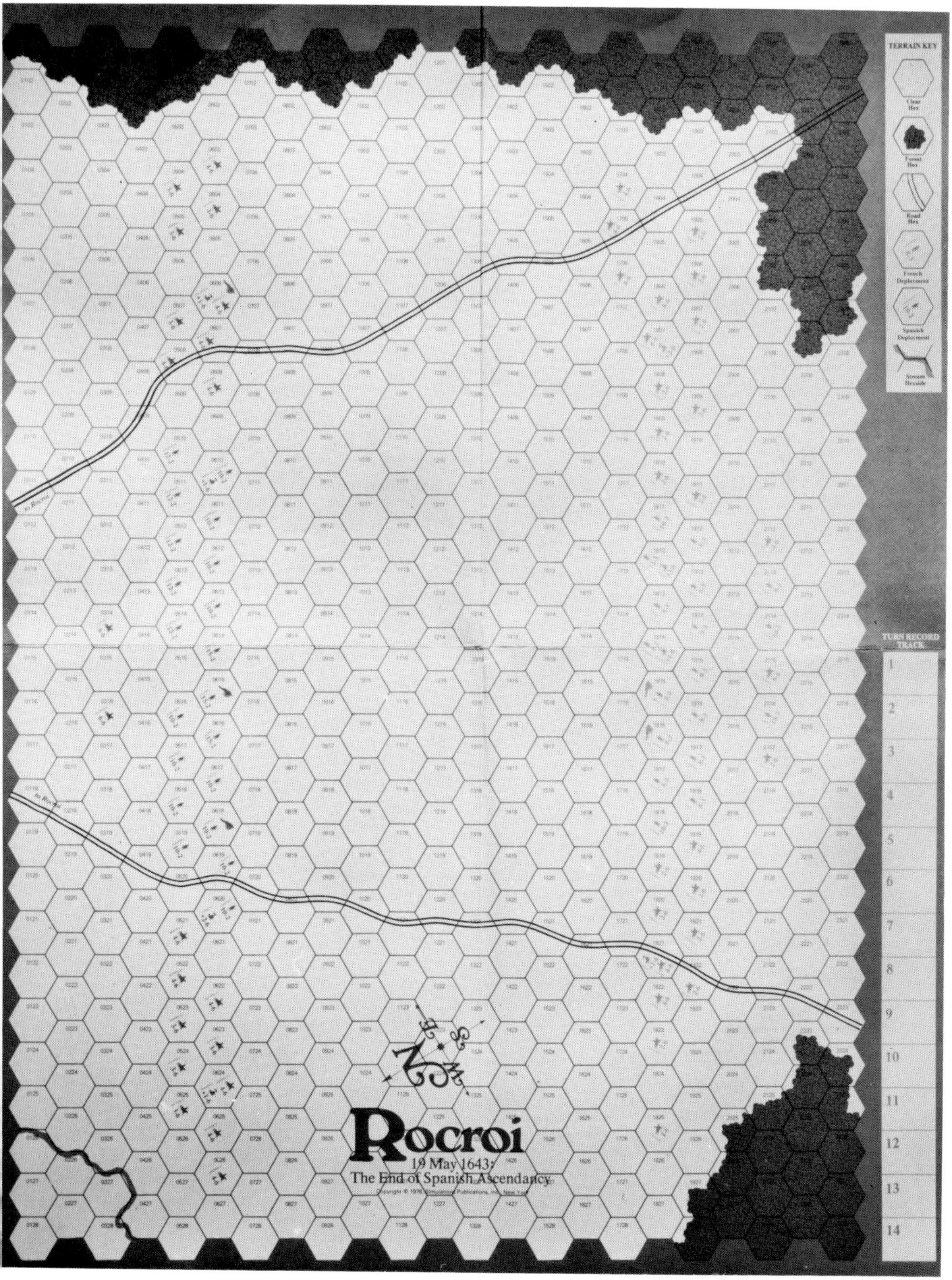

3 The mapboard of *Rocroi.*

gon grid system. It suggests a mathematical flavour, and they begin to wonder if they are going to be asked to calculate cosines or differential equations. There is no need to fear anything like this once the basic purpose of hexagons is understood. We need *some* sort of grid to make movement easy; instead of measuring distances (as in miniature games) we can now just count hexagons, allowing for terrain effects on speed (e.g. most things move slowly through mountains). In the infancy of wargaming, maps with square grids were tried, as in chess. But this leaves the question of what to do about the corners. If one can cross diagonally over corners then the squares are *de facto* changed into eight-sided figures (four sides and four corners); if one cannot, then everything zigzags bizarrely over the map in a series of straight lines and right-angled turns, giving a thoroughly odd effect.

The terrain is handled in exactly the same way as on road maps: different colours and signs indicate what is to be found in each hex. There may be several features in and around one hex, e.g. a fortress on a mountain behind a river, or there may be nothing shown, which means that it is nondescript flat ground (or sea/air in the non-land games).

THE UNITS

Each unit in the battle is represented by a square cardboard counter (see the illustration in Chapter 1 for a variety of units) of a size to fit in the map hexes, and mounted for better durability. The counter will usually have everything you need to know about the unit printed on it: size, type, combat strength, speed and (usually indicated by the colour) which side it is on. We shall examine some of these more closely in a minute, but first note that the designer has a wide choice as to what sort of units he uses. Say the game features the German assault on the Soviet Union. Should there be a unit for each army? Each division, like *Drang Nach Osten*? Each man? (Don't be ridiculous!) His decision will be influenced by the map. Players like to have forces which fit comfortably on the board without piling ten counters onto a hex. On the other hand, too few counters and they will wander around the map without much opportunity for joint action, making a boring game, or most of the board will go unused (a common flaw in early designs). Moreover, if a big-unit (e.g. corps) level is chosen, then the units would fight over a front scores of miles in width, so the hexes should also be on this scale. In this, as in most aspects of wargaming, there is something for every taste, with games like *Sniper* going right down to individual soldiers in Second World War street-fighting.

The most important information on the counter, apart from the nationality, is usually expressed in two or three figures at the bottom, saying something like 6-3 or 6-4-3.

The first figure is normally the *combat factor*; if there are three figures, then the first two will both be combat factors, giving attack and defence strength respectively, but otherwise the attack and defence strengths are identical. The last figure is the *movement factor*, or speed.

We shall see later how the combat factors are used. The movement factor is simply used in conjunction with the hex grid and terrain modifications described earlier. If the terrain is not of the kind that affects movement, you can simply move the unit up to the number of hexes shown as its movement factor: a 6-3 could move three hexes from its current position. If the unit is motorized and on a road it may be able to move much faster: the terrain chart might say that road movement only 'costs' one fifth movement points per hex, so a 6-3 could move fifteen hexes along a road, or ten hexes along the road, and then one hex off the road. Alternatively, the terrain may be rough, or steep, slowing speed to perhaps just a single hex per turn until the open is reached again.

All units move every turn (unlike e.g. chess). In some games, certain units move twice each turn, usually to allow armour to exploit a hole in the enemy line after combat. A *turn* represents a space of time, usually long in big-unit games (perhaps three months, to allow a frontal offensive each turn) and short in low-level games, typically an hour or so. The game will usually start and finish at fixed turns, so that players will know the maximum number of turns they may play and the game cannot go on forever; fifteen turns is probably about average, but 200-turn games have been known, and a couple of games last just two turns

In some games, the players move simultaneously, but to avoid the increase in paperwork which this necessitates (as neither must then see what the other is up to) it is more usual to alternate, as in chess, which need not be very unrealistic if the amount that can be done before the other side reacts is limited.

The prudent purchaser will have a close look at the number of turns. If there are over twenty, it is probably a long game, unless many of the turns are very quickly made. Anything over 100 turns is best suited to shipwrecked mariners without hope of rescue for a year or two, though there may be shorter versions, or it may be such a good game that one is prepared to live with the fact that it cannot be finished in an evening's play.

In most games, one can have more than one unit on a hex, providing they are on the same side (this is called *stacking*), but there are nearly always limits, dictated by real-life practical considerations: three armies quartered in Little Wopping, normal population 136, cannot expect to fight together with their customary effectiveness. This is a relief for the player, who does not want to fiddle with great stacks of counters. The usual limit is two or three units per hex, with modifications when different-sized formations (battalions and regiments, for example) are present in the same game.

Units on opposite sides are normally forbidden to occupy the same hex at once. If you want to take Little Wopping and your opponent has a unit there already, you must move up outside the village and force him out in combat before you can move in.

COMBAT

When enemy units get within shooting range of each other, fighting will naturally tend to break out; in some games this is compulsory, in others up to either or both of the players to decide. The following straightforward system is used in nearly every game when one player's units attack those of his opponent.

The combat factors (using the appropriate one of attack and defence factors when these are separate) of the units on each side are added together, and the attacker's total is divided by the defender's (rounding in favour of the defender) to get a simple ratio like 1–2 or 7–1. Thus if three regiments marked 6-3 and a full division marked 15-2 attack a pile of three defending 7-4s, we disregard the movement factors and add the combat factors:

Attacker: $3 \times 6 = 18$
plus $1 \times 15 = 15$
Total: 33.
Defender: $3 \times 7 = 21$.

33 divided by 21 rounds down to 1–1, so the attacker could in fact equally well have used two of his regiments elsewhere, and got 21–21 odds, or still 1–1, though there may be other advantages in the greater force. If, however, the attacker had brought in two more regiments with combat strength six, then he could have brought the odds up to 45–21, which rounds down to 2–1.

Having discovered the ratio of strength, we consult the *Combat Results Table* (CRT) which comes with every game; the illustration is the CRT from the Avalon Hill 'classic' series which includes *Stalingrad*, *D-Day* and *Waterloo*. A die is thrown to decide which of the six outcomes in the column for battles at this ratio will occur. In our

COMBAT RESULTS TABLE

ODDS / DIE ROLL	1—6	1—5	1—4	1—3	1—2	1—1	2—1	3—1	4—1	5—1	6—1	ODDS / DIE ROLL
1	A elim	A elim	A back 2	A back 2	D back 2	D elim	D elim	D elim	D elim	D elim	D elim	1
2	A elim	A elim	A elim	A back 2	Exchange	Exchange	Exchange	Exchange	Exchange	D back 2	D back 2	2
3	A back 2	A back 2	A back 2	A back 2	A back 2	D back 2	D back 2	D back 2	D elim	D elim	D elim	3
4	A elim	A back 2	A back 2	A back 2	A back 2	A back 2	A back 2	D back 2	D back 2	D back 2	D elim	4
5	A elim	A elim	A elim	A elim	A elim	A elim	Exchange	Exchange	D back 2	D elim	D elim	5
6	A elim	A elim	A elim	A elim	A elim	A elim	A elim	D elim	D elim	D elim	D elim	6

Odds greater than 6 to 1 or 1 to 6 mean automatic elimination.

HOW TO RESOLVE COMBAT

STEP 1: Refer to the Chart *below* to reduce battle odds to a basic odds comparison shown on the Combat Results Table. To do this simply cross-index the attacker's factor (vertical column) with the defender's factor (horizontal line). For example, battle odds of 30 to 9 breaks down to 3 to 1. (When a dash (—) appears the weaker Units are automatically eliminated.)

STEP 2: The Die is rolled once by the attacker.

STEP 3: The Die roll is matched up with the basic odds comparison to get the result of combat. For example, at 3 to 1 a Die roll of 6 means that all defending Units are eliminated. Repeat STEPS 1 through 3 for each separate battle.

DEFENDER'S FACTOR

OR	1	2	3	4	5	6	7	8	9	10	11	12	13	14	15	16	17	18	19	20
1	1-1	1-2	1-3	1-4	1-5	1-6	—	—	—	—	—	—	—	—	—	—	—	—	—	—
2	2-1	1-1	1-2	1-2	1-3	1-3	1-4	1-4	1-5	1-5	1-6	1-6	—	—	—	—	—	—	—	—
3	3-1	1-1	1-1	1-2	1-2	1-2	1-3	1-3	1-3	1-4	1-4	1-4	1-5	1-5	1-5	1-6	1-6	1-6	—	—
4	4-1	2-1	1-1	1-1	1-2	1-2	1-2	1-2	1-3	1-3	1-3	1-3	1-4	1-4	1-4	1-4	1-5	1-5	1-5	1-5
5	5-1	2-1	1-1	1-1	1-1	1-2	1-2	1-2	1-2	1-2	1-3	1-3	1-3	1-3	1-3	1-4	1-4	1-4	1-4	1-4
6	6-1	3-1	2-1	1-1	1-1	1-1	1-2	1-2	1-2	1-2	1-2	1-2	1-3	1-3	1-3	1-3	1-3	1-3	1-4	1-4
7	—	3-1	2-1	1-1	1-1	1-1	1-1	1-2	1-2	1-2	1-2	1-2	1-2	1-2	1-3	1-3	1-3	1-3	1-3	1-3
8	—	4-1	2-1	2-1	1-1	1-1	1-1	1-1	1-2	1-2	1-2	1-2	1-2	1-2	1-2	1-2	1-3	1-3	1-3	1-3
9	—	4-1	3-1	2-1	1-1	1-1	1-1	1-1	1-1	1-2	1-2	1-2	1-2	1-2	1-2	1-2	1-2	1-2	1-3	1-3
10	—	5-1	3-1	2-1	2-1	1-1	1-1	1-1	1-1	1-1	1-2	1-2	1-2	1-2	1-2	1-2	1-2	1-2	1-2	1-2
11	—	5-1	3-1	2-1	2-1	1-1	1-1	1-1	1-1	1-1	1-1	1-2	1-2	1-2	1-2	1-2	1-2	1-2	1-2	1-2
12	—	6-1	4-1	3-1	2-1	2-1	1-1	1-1	1-1	1-1	1-1	1-1	1-2	1-2	1-2	1-2	1-2	1-2	1-2	1-2
13	—	6-1	4-1	3-1	2-1	2-1	1-1	1-1	1-1	1-1	1-1	1-1	1-1	1-2	1-2	1-2	1-2	1-2	1-2	1-2
14	—	—	4-1	3-1	2-1	2-1	2-1	1-1	1-1	1-1	1-1	1-1	1-1	1-1	1-2	1-2	1-2	1-2	1-2	1-2
15	—	—	5-1	3-1	3-1	2-1	2-1	1-1	1-1	1-1	1-1	1-1	1-1	1-1	1-1	1-2	1-2	1-2	1-2	1-2
16	—	—	5-1	4-1	3-1	2-1	2-1	2-1	1-1	1-1	1-1	1-1	1-1	1-1	1-1	1-1	1-2	1-2	1-2	1-2
17	—	—	5-1	4-1	3-1	2-1	2-1	2-1	1-1	1-1	1-1	1-1	1-1	1-1	1-1	1-1	1-1	1-2	1-2	1-2
18	—	—	6-1	4-1	3-1	3-1	2-1	2-1	2-1	1-1	1-1	1-1	1-1	1-1	1-1	1-1	1-1	1-1	1-2	1-2
19	—	—	6-1	4-1	3-1	3-1	2-1	2-1	2-1	1-1	1-1	1-1	1-1	1-1	1-1	1-1	1-1	1-1	1-1	1-2
20	—	—	6-1	5-1	4-1	3-1	2-1	2-1	2-1	2-1	1-1	1-1	1-1	1-1	1-1	1-1	1-1	1-1	1-1	1-1
21	—	—	—	5-1	4-1	3-1	3-1	2-1	2-1	2-1	1-1	1-1	1-1	1-1	1-1	1-1	1-1	1-1	1-1	1-1
22	—	—	—	5-1	4-1	3-1	3-1	2-1	2-1	2-1	2-1	1-1	1-1	1-1	1-1	1-1	1-1	1-1	1-1	1-1
23	—	—	—	5-1	4-1	3-1	3-1	2-1	2-1	2-1	2-1	1-1	1-1	1-1	1-1	1-1	1-1	1-1	1-1	1-1

OR	21	22	23	24	25	26	27	28	29	30	31	32	33	34	35	36	37	38	39	40
1	—	—	—	—	—	—	—	—	—	—	—	—	—	—	—	—	—	—	—	—
2	—	—	—	—	—	—	—	—	—	—	—	—	—	—	—	—	—	—	—	—
3	—	—	—	—	—	—	—	—	—	—	—	—	—	—	—	—	—	—	—	—
4	1-6	1-6	1-6	1-6	—	—	—	—	—	—	—	—	—	—	—	—	—	—	—	—
5	1-5	1-5	1-5	1-5	1-5	1-6	1-6	1-6	1-6	1-6	—	—	—	—	—	—	—	—	—	—
6	1-4	1-4	1-4	1-4	1-5	1-5	1-5	1-5	1-5	1-5	1-6	1-6	1-6	1-6	1-6	1-6	—	—	—	—
7	1-3	1-4	1-4	1-4	1-4	1-4	1-4	1-4	1-5	1-5	1-5	1-5	1-5	1-5	1-5	1-6	1-6	1-6	1-6	1-6
8	1-3	1-3	1-3	1-3	1-4	1-4	1-4	1-4	1-4	1-4	1-4	1-4	1-5	1-5	1-5	1-5	1-5	1-5	1-5	1-5
9	1-3	1-3	1-3	1-3	1-3	1-3	1-3	1-4	1-4	1-4	1-4	1-4	1-4	1-4	1-4	1-4	1-5	1-5	1-5	1-5
10	1-3	1-3	1-3	1-3	1-3	1-3	1-3	1-3	1-3	1-3	1-4	1-4	1-4	1-4	1-4	1-4	1-4	1-4	1-4	1-4
11	1-2	1-2	1-3	1-3	1-3	1-3	1-3	1-3	1-3	1-3	1-3	1-3	1-3	1-4	1-4	1-4	1-4	1-4	1-4	1-4
12	1-2	1-2	1-2	1-2	1-3	1-3	1-3	1-3	1-3	1-3	1-3	1-3	1-3	1-3	1-3	1-3	1-4	1-4	1-4	1-4
13	1-2	1-2	1-2	1-2	1-2	1-2	1-3	1-3	1-3	1-3	1-3	1-3	1-3	1-3	1-3	1-3	1-3	1-3	1-3	1-4
14	1-2	1-2	1-2	1-2	1-2	1-2	1-2	1-2	1-3	1-3	1-3	1-3	1-3	1-3	1-3	1-3	1-3	1-3	1-3	1-3
15	1-2	1-2	1-2	1-2	1-2	1-2	1-2	1-2	1-2	1-2	1-3	1-3	1-3	1-3	1-3	1-3	1-3	1-3	1-3	1-3
16	1-2	1-2	1-2	1-2	1-2	1-2	1-2	1-2	1-2	1-2	1-2	1-2	1-3	1-3	1-3	1-3	1-3	1-3	1-3	1-3
17	1-2	1-2	1-2	1-2	1-2	1-2	1-2	1-2	1-2	1-2	1-2	1-2	1-2	1-2	1-3	1-3	1-3	1-3	1-3	1-3
18	1-2	1-2	1-2	1-2	1-2	1-2	1-2	1-2	1-2	1-2	1-2	1-2	1-2	1-2	1-2	1-2	1-3	1-3	1-3	1-3
19	1-2	1-2	1-2	1-2	1-2	1-2	1-2	1-2	1-2	1-2	1-2	1-2	1-2	1-2	1-2	1-2	1-2	1-2	1-3	1-3
20	1-2	1-2	1-2	1-2	1-2	1-2	1-2	1-2	1-2	1-2	1-2	1-2	1-2	1-2	1-2	1-2	1-2	1-2	1-2	1-2
21	1-1	1-2	1-2	1-2	1-2	1-2	1-2	1-2	1-2	1-2	1-2	1-2	1-2	1-2	1-2	1-2	1-2	1-2	1-2	1-2
22	1-1	1-1	1-2	1-2	1-2	1-2	1-2	1-2	1-2	1-2	1-2	1-2	1-2	1-2	1-2	1-2	1-2	1-2	1-2	1-2
23	1-1	1-1	1-1	1-2	1-2	1-2	1-2	1-2	1-2	1-2	1-2	1-2	1-2	1-2	1-2	1-2	1-2	1-2	1-2	1-2

4 The Combat Results Table of the 'classics'.

case of a 1–1 battle, a 3 or 4 would compel a two-hex retreat for the defenders and the attackers respectively. A 1 would result in the elimination of the defence, while a 5 or 6 would result in the attacking units being destroyed and removed from the board. A 2 causes an 'exchange' in which the defenders would remove all their units and the attackers at least as many combat factors. This often leaves them with some, if they started the attack with the larger force. In the example, the two surplus regiments would be left, perhaps making the player glad he had not put them somewhere else after all.

This chance element reflects reality much more closely than the certitudes of, say, chess. No general can ever be sure of a reasonably even battle resulting in a particular outcome, although he can be pretty confident that some results will not occur. In the game using the CRT shown, attackers with favourable odds of 3–1 can be certain that they will not be wiped out by the defenders, although they might take substantial losses (exchange), force the position but let the defenders escape (defender back 2) or succeed in wiping out the defence (defender eliminated). Only if a 7–1 disparity of forces can be amassed is the result a foregone conclusion.

However, the CRT in the 'classics' is one of the 'bloodiest' in use, and as such gives rather crude swings of fortune, especially at lower odds than 3–1. More recent designs tend to favour more retreat or partial loss results, instead of the blockbuster 'eliminated' outcome, which is such a potential disaster that 'classic' attacks under 3–1 are rare. In games using a CRT dominated by retreat results, surrounding the enemy becomes of crucial importance, as, if he is unable to retreat as a result, then he is eliminated instead; in real life a unit surrounded and unable to hold its ground would indeed surrender or be wiped out.

The uncertainty about the exact outcome of each battle presents the player with some fascinating problems. Instead of being able to calculate every aspect of the situation several moves ahead, like a chess player, he is forced to keep his plans sufficiently flexible to allow for a range of possible positions after the battle. An offensive which may look fine when the most likely results are assumed may lead to disaster if one or two key battles go the wrong way. This situation is one of the best litmus-tests for knowledge of the game. The experienced player will attempt to cover every outcome, and will be disappointed but not horrified by the worst combat results. The beginner will often opt for a more spectacular assault; if the result is catastrophic he will then complain bitterly about his rotten luck. This is not to say that risks should not be taken at times; if the chances are good then daring will pay off in the long term. The vital thing is to minimize the risk of throwing the whole game away with a madcap assault, unless it is the only remaining chance of victory.

Terrain can affect combat strength. Most forms of irregular terrain benefit the defender, as he can direct his fire from prepared positions, while the attacker has to struggle through the undergrowth, stumble up a mountain side, or ford a river. The usual effect of such terrain is to double the strength of the defender, although in the case of rivers this is cancelled if some of the attackers have already crossed, as they are assumed to be able to pin down the defence while their brethren cross over to join them. In some games it is possible to be tripled on defence, usually in specially constructed fortresses, but defence bonuses are rarely allowed to reinforce each other – the attacker may be struggling with prickly bushes, or in danger of being swept downriver by a powerful current, but hardly both at once!

ZONES OF CONTROL; RETREATS; ADVANCES

As we have noted, most units are designed to fit physically into a single hex. It is a cardinal rule of nearly all games that units cannot enter enemy-occupied hexes without first destroying the enemy; almost the only exception to this is that vastly superior forces are sometimes allowed to destroy the defenders *as* they pass through, so that a knot of home guards cannot expect to hold up a division of tanks. However, units usually have an effect on the six hexes immediately adjacent to them, called their *zone of control* (ZOC). In older games and many recent designs (especially on modern warfare), enemy units entering a ZOC have to stop and fight. Other games allow very fast units to slip through a ZOC after a movement delay, to simulate armoured penetration of a thin defence line, while games based on early (pre-nineteenth century) warfare sometimes eliminate the rule altogether, since it is really modern weaponry which enables units to exert an influence on neighbouring areas – it wouldn't work with cutlasses.

ZOCs also work to prevent advances and retreats, and this makes it easier to destroy an enemy unit by blocking his retreat routes. The older games do not even allow the retreating unit to join adjacent friendly forces if those units are in an enemy ZOC; most games nowadays allow this, and also allow supply lines to be traced through friendly units even if these are in enemy ZOCs.

We have not previously dealt with advances. There are two systems for dealing with the momentum gained by successful attackers. The 'classic' games with the CRT which we saw earlier allow attackers which have successfully stormed a position where the defence was *doubled* to occupy that position. This is based on the assumption that the battle will in this case probably have hinged on control of the defensive strongpoint, so if an attack has succeeded it must be because the strongpoint has been captured. Other games include advances as a result of victory on the CRT itself, which has a similar effect, except that one may then get a longer advance, and into undoubled terrain. This is more realistic, reflecting the various degrees of resistance put up by the defence.

The alternative approach is the system referred to earlier, in which some or all units are allowed to move a second time after combat, thereby exploiting any holes which have appeared in the enemy line. When this technique can be used, defence lines need to have a back-up cordon of units to be secure from major penetration, encirclement, and other horrors of that sort. Quite often an advance will then leave the leading units exposed to a shattering counter-attack from several sides on the enemy turn.

VICTORY CONDITIONS

If you are playing a game for the first time, the very first thing to do after a glance at the map is to read the victory conditions. These are laid down in the rules with two things in mind: historical accuracy and a balanced game. These aims sometimes conflict. Thus, in *Sinai*, 1956 Scenario, the Arab player has done magnificently if he has limited the scale of the Israeli victory, but in historical terms he has still lost. The victory conditions in the game will usually relate to game balance, and say what a 'good result' for each player would be, in view of the situation. There will often be a note about *how* good a result would have been needed historically to pull off an actual victory.

Victory conditions naturally depend on the scope of the game. If we are simulating the Second World War, then the conditions will probably relate to the amount of area or the number of key centres held on a particular date; if Germany holds Moscow or the USA occupies Berlin in 1945, then the occupier has almost certainly won (failing some disaster at home). If we are looking at the Battle of Jutland, then it is more likely to be the number and size of ships surviving which counts, rather than a territorial objective. If the game deals with a tactical skirmish, then the victory conditions may relate to the number of units left on a particular hill at the end of the day. But whether you are supposed to end up controlling two thirds of the galaxy or the village town hall, it is crucially important that you know what you are aiming for and keep it constantly in mind. Even experienced players sometimes devise strategies which lead to all kinds of great victories except those which they were supposed to achieve, and consequently lose to their infuriatingly smug opponents who have read the rules better: true, they only have the one platoon left in the ammunition dump, and it is facing imminent attack by a whole battalion, but the victory condition was holding the dump for twelve turns and they have succeeded! (The condition no doubt assumes that large-scale reinforcements would have arrived on the thirteenth turn.)

Game designers may not always realize the extent to which the victory conditions and the scale of the game can affect players' enjoyment. A good victory condition from this point of view will be something of self-evident importance in the real-life situation shown. Thus the Japanese in *Midway* need to capture Midway Island rapidly, as well as obtaining a favourable balance of losses; we can immediately see that this would indeed have left them in a much more powerful position than before the battle. In some of the games with territorial objectives, however, the victory condition is merely seizing a certain number of these; in *Third Reich*, for instance, it would have made a great deal of difference in real life whether the objectives held by Germany were in Russia or in England, but victory is achieved by a simple count. The rationale of this approach is that it keeps things simple: players want to be able to see at a glance how they are getting on, and not have to calculate four or five different factors.

Whether one prefers a small-scale game with tactical objectives or a large-scale one with strategic targets is very much a matter of taste. There is a certain grandeur about fighting to win the Second World War or struggling with Napoleon for mastery of Europe which is missing from tactical games with a goal of possession of Hill 128 or occupation of some key village. On the other hand, it is possible to achieve marvellous detail on the tactical level, with every platoon and heavy weapon having its own counter and function. In the games of grand strategy the idea is to focus on the wide-ranging decisions and make the individual battles fairly simplified; this can mean, as your space ships sweep over the universe or your armies roll into Germany, that you have slight doubts over whether it would really be like this in practice! In between the extremes there are a number of different levels, the most popular being the 'operational', which normally deals with an individual battle, e.g. *Battle of the Bulge*, *Sinai*, and *Jerusalem*. Operational-level games combine some of the virtues of each of the other types: they hinge on the actions of individual regiments and divisions, but they have strategic goals and usually simulate some important battle; most people find it fun to compare the game outcome with what really happened.

SUPPLY

We will conclude this chapter on the basic concepts of wargaming with a feature which is rather less basic than the others: supply rules. Many of the older games do not, in fact, have supply rules, or only in an elementary way (e.g. by having units cut off from their lines eventually surrender). However, it is now unusual in non-tactical games to have no provision for supply considerations.

The concept is usually handled in two ways: either units must be able to trace a line free of enemy zones of control to a road or rail line leading to a supply source like a large town or home territory, or there are actual supply units, which are often used up if they support an attack or group of attacks. The first system makes play a good deal easier than the second, but tends to be unrealistic if the game involves a severe supply problem, as for instance after an invasion. The second method makes the exact placement of the supply units highly important, particularly as one supply unit can support several attacks before it is used up, so long as they take place simultaneously in the same area. (I have never understood the logic of this standard rule, but it does make things simpler than a strict one-attack-per-supply-unit rule, which would necessitate a vast number of supply units cluttering up the map.)

Supply often plays an important role in the victory conditions. When some objective is to be seized, it is undesirable to have this done by some trifling little unit which has slipped through the lines, and is now miles away from the main force with a seething mass of enemies in between. It is, therefore, normal to require that the unit occupying the objective be in supply.

The supply rules frequently distinguish between supply for attacks and supply for defence, or for just moving about without combat. The supply cost of attacks is obviously much greater, and supply units are expended only when used for offensive purposes.

PART I

Strategy

Chess-players are often taught to play first with a piece or two, in small-scale tactical exercises, before they are gradually allowed to advance to the glories of the full game with the complete 'armies' present. This has two results: first, many *never* acquire a strategic flair, being too occupied with looking for the ingenious little combinations they learned early on; secondly, the purpose of the tactical manœuvres is often obscure, because the centre-seizing or pawn-front-busting which they are to accomplish is not taught until later. As far as possible, we shall do it the other way around.

In this part, we look at the planning of strategy. While most players become adequate tacticians (through learning from someone, or by bitter experience), there are many who never show more than the most elementary grasp of strategy. Because this does not necessarily mean that individual battles are badly planned, this is not immediately obvious. It is hard to pinpoint any specific decisive error, but somehow things seem to slide gently downhill. A striking example of this in a *Stalingrad* game appears in Chapter 3.

After this general chapter on strategic thinking, Chapter 4 deals with the question of reserves and build-up: too many reserves, and available forces stand unused; too few and the line becomes vulnerable to penetration at isolated points. The commitment of units has to be planned ahead to combat this problem effectively, and the need for reserves depends on the rules for movement and the nature of the Combat Results Table.

Chapter 5 relates to the advanced type of game in which players have control over political and economic decisions in addition to purely military actions. Advanced planning here may need to stretch over several game-years, and good timing is often vital.

It should be noted that the word 'strategic' is relative to the scale of the game. A strategic plan in *World War III* will deal with the time-scale for capturing continents, but a strategy for *Sniper* will have more to do with the order in which individual buildings are assaulted. *Every* game has its strategic and tactical aspects.

3 THE FOREST AND THE TREES

When you sit down to a game, you will certainly have some kind of strategy in mind. You cannot really help doing so, unless you have forgotten to read the victory conditions. They will help fix in your mind the general way you want the battle to go: a high enemy casualty rate, an early seizure of a key central point, or whatever the conditions may be. Many players leave it at that, behaving like bicyclists on a steep hill: they glance up from time to time to make sure they are not going to run into anything nasty, but otherwise concentrate on hard pedalling. The players of this type bash ahead on all fronts, taking every tactical opportunity that appears and only bearing the victory conditions in mind as a general check on how things are going.

This is not always harmful, as sometimes a purely tactical approach in the right general direction will be all that is required. More usually, it will land the player in the soup about half way through. Very likely it will all look fine for a while, with careful tactical placement on each front leading to a steady series of net gains. If we are to use the saying about not seeing the forest for the trees, then the 'trees' (the local battles) individually may be genuinely ideal, given the chosen plan (or absence of one). The general situation, however, is likely to be deteriorating steadily, as the different sectors get out of phase with each other, key units go chasing up blind alleys in ingenious but short-sighted tactical thrusts, and the overall strategy is changed to fit with the openings available on each turn. Gradually, a competent opponent will notice what is happening, and encourage the process by putting tempting targets for the other player in positions which force a further strategic dislocation to get at them. The player will probably not even realize what has gone wrong, but merely note that 'unfortunately' he has not got his units where he wants them towards the end, 'despite' his spectacular victories earlier on.

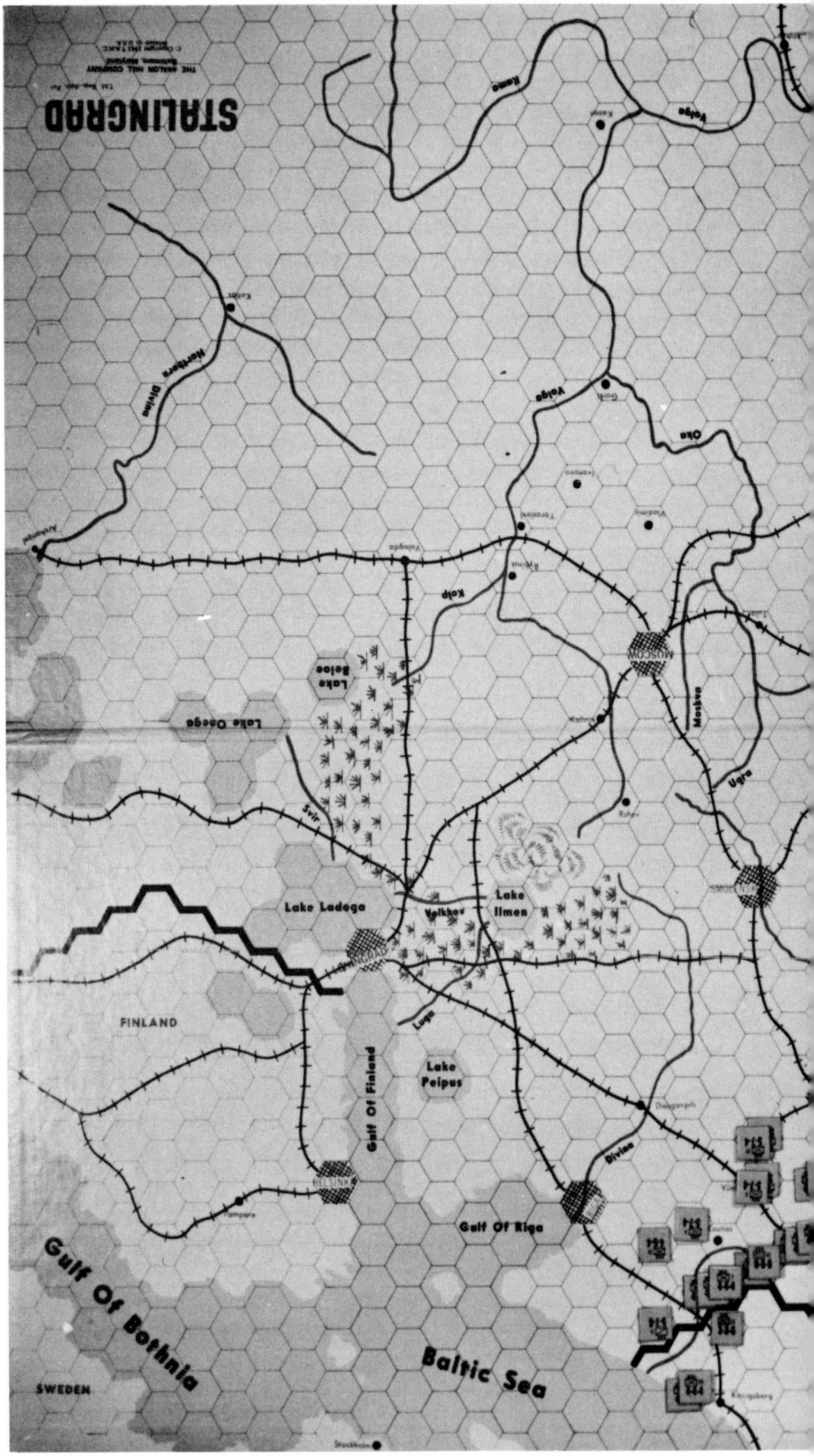

STALINGRAD
Kama
Volga
Northern Dvina
Oka
Kolp
Lake Beloe
Lake Onega
Moskva
Ugra
Svir
Lake Ladoga
Volkhov
Lake Ilmen
FINLAND
Luga
Gulf Of Finland
Lake Peipus
Dvina
Gulf Of Riga
Gulf Of Bothnia
Baltic Sea
SWEDEN

5 Mapboard of *Stalingrad* showing a position at the start of Germany's December 1941 turn.

'What an unbalanced game!' he will exclaim irritably, and consign it to the nether depths of his cupboard.

To take a concrete example, we can look at the AH 'classic' game *Stalingrad*, which despite its name deals with the whole Second World War East Front, and is notable for its extremely clear-cut strategic choices. Illustration 5 shows a position in the game. The victory condition for Germany is the simultaneous possession for two turns of Leningrad, Moscow and Stalingrad; if the Russians avoid this by May 1943 (after twenty-four turns simulating one month each) then they win (as it is assumed that the Germans would by then have passed their peak, as they had historically). The main geographical features, apart from the three key cities, are a number of rivers (which give the defender doubled strength), some subsidiary cities (which again double defence strength) and, crucially important, the Pripyat marshes which divide the front in the middle. The marshes do not increase defensive strength, but they can only be traversed one hex at a time. Digressing briefly, this is one case where a more recent game than a 'classic' would draw a distinction between different kinds of unit, armoured units being particularly frustrated by marshland.

The Pripyat marshes can make life difficult for the Russian player, as he has to decide which side of them each of his defenders will operate. The German player, however, is staring into the jaws of a much more subtle trap. For the first few months, it will hardly matter at all where he commits his forces on the central fronts, since the rail lines running near the border allow him to shuttle his troops to and fro as the situation demands; *Stalingrad* units can move by rail for up to ten hexes without using up their movement factors. Nor is it of crucial importance where early progress is made, as the three target cities are scattered over the board; sooner or later the German needs to break through in all sectors. The tactical approach described earlier will therefore yield very satisfactory dividends for a while.

Illustration 5 shows the position after six months of this, at the start of Germany's December 1941 turn. Experience in the game shows that Germany needs to win or get very close to victory in 1942 or have no chance of success, as from May in that year the Soviet replacement rate is at full power as the economy weathers the shock of the invasion.

All movement (normal and rail) in the winter turns is at half speed, hampering the rapid advance needed. Nevertheless, the German position superficially looks not bad, especially after a glance at the casualty list: the Soviet forces have lost fifteen units with 83 defence and 56 attack factors, compared with German losses of nine units with 20 defence and attack factors (German units have the same combat factors for attack and defence, being better trained for attack than the Soviet troops at this stage of the war), not counting six small Finnish units with 15 factors which the Soviet player has wiped out. On the map, the Germans have made long strides in the south, and are glaring at the Russians over the Dnepr river, the last natural barrier before Stalingrad and the Moscow area. Progress in the north has been slow, but two river lines have been broken, and the Russian defences are very thin: only eight of the remaining nineteen Russian units have been committed here, and all of them have 6-7 defence factors, with three bigger units (9-10 defence factors) all assigned to the Dnepr. The Russian replacement rate is 15 defence factors per turn, rising to 18 in May 1942 (historical accuracy would dictate 18 and 24 respectively, but AH recommends the lower rate to give a more balanced game), while the Germans only get 4 DF/AF per turn. The Soviet replacements are cut by a third, however, for each of the target cities captured by Germany, and once the Russians are pushed away from their defensive lines into the open, their losses will become very large. The entire German armoured spearhead has survived, to help this process along.

The tactically oriented German player will no doubt be feeling rather pleased with himself. So, however, will his opponent, because the Soviet losses have gone in an excellent cause: luring the German forces into such a mess that they are going to have to waste virtually the whole winter repositioning. Strategically, the Axis position is quite gruesome.

The problem centres on a single point: the armoured units, with their 6-8 attack factors each, have been split up. Two 8-8-6s are with the northern units, while the rest are in the south, mostly concentrated at the furthest point of advance. *Stalingrad* uses the classic Combat Results Table shown in Chapter 2, and as noted there one needs a 3-1 attacking edge to be sure of avoiding the elimination of the entire attacking force (a total catastrophe if 50 factors or so are involved). Three *Stalingrad* units can be stacked on a hex, but units have to stop when they enter an enemy zone of control, by moving next to an enemy piece. The problem facing the German player becomes clear when we consider how, for instance, he is to get a 3–1 attack on the 5-7-4 (14 Corps) in Minsk, just north of the Pripyat marshes. This has defence factor 7, and cities double defence, so it defends with strength 14. To get 3–1 the German needs to get 42 factors. Now, he has lots of units in the vicinity – 70 factors, in fact! But Minsk can only be attacked from the two hexes in front of the city: if the German tries to slip round to the north, he has to stop and fight the four Russian 5-7-4s northwest of Minsk; if he tries to get round to the south, it will take three turns sloshing through the Pripyat, at one hex per turn, by which time the Soviet player will be able to bring up reinforcements, and in any case will have gained valuable time.

But two hexes in front of Minsk will only allow six attacking units, because of the stacking limit. We now see why it is important to have powerful units. The two 8-8-6s contribute 16 attack factors, but all the other forces in the area are 4-4-4s or even weaker; four of these contribute another 16 attack factors, making the odds:

Germany: $2 \times 8 = 16$ AF	*Soviet Union:* 7DF
$4 \times 4 = 16$ AF	Minsk: doubles
Total 32 AF	Total 14 DF
32:14=2–1	

A 2–1 is not to be considered except in the direst emergency, for capturing Minsk would not win the game, while losing 32 German factors in one battle would certainly lose it.

The tactical player therefore curses his 'bad luck' at having the big units too far away to take Minsk. But similar considerations make it impossible to get a 3–1 anywhere on the entire southern front either, those two 8-8-6s in the north being sorely missed. The only place where a 'safe' attack is possible is against the blocking units north-west of Minsk. The German should certainly make this, but while he may (failing a counter-attack) force the block back slightly, widening the angle of possible attack on Minsk, the Russian can reinforce Minsk with another 5-7-4, and this will hold it safe for another couple of turns, despite a technique called the 'soak-off' (described in Chapter 6) which enables concentration on one defending unit at a time.

It is therefore extremely urgent to regroup the armour together so that they can break the defensive lines. Unfortunately, it is winter, and movement is at half speed. The 8-8-6s in the north, for instance, can move three hexes, and then a further five by rail. A simple count shows that at this rate it will take three months to reach the southern front! Conversely, most of the southern armour will take two or three months to get to the north. During three turns, the Russians will have brought in 45 defence factors of reinforcements to the German 12, and the combined armour will not be enough to make up for the delay. Far from the position looking promising for the Germans, as the tactical expert fondly believed, it is almost resignable. All he can do is fling his units at under 3–1 odds against the various doubled positions (1–2 attacks are recommended here, giving a one-sixth chance of forcing the position with a 'defender back 2' but risking *relatively* little) and hope that remarkable luck with the dice will make up for his lack of strategic sense. It is very unlikely that he will succeed.

What the German did wrong was to allow himself to be distracted from a coherent strategy by the chance of easy victories in both north and south over individual Soviet units. It is not enough to be 'going in the right direction'; he should have known that he would need an armoured concentration to break the defensive line in the south, and positioned his *panzers* there before the snows set in, leaving a strong infantry force in the Minsk area to erode the Russian blocking positions and work through the Pripyat. Alternatively, he could have committed his armour in the north, using small infantry units to try and break the Dnepr with probes at 1–2 odds. The first plan is probably preferable, but either way the campaign would not grind to a halt throughout the winter. Moreover, if the armour had been massed in the south and had successfully broken the Dnepr defences, then the road to Stalingrad and Moscow would have been over mostly open country, leaving the Russian with the unenviable choice of fighting without defensive bonus against a numerically superior enemy, or allowing the Germans to reach two of the cities and slash the vital replacement rate. The Russians would still have a fighting chance, because with the weak northern advance and the defeat of their Finnish allies the Germans would have difficulty in taking Leningrad; *Stalingrad* is in fact generally believed to give an advantage to the Soviet player with best play on both sides, unless the replacement rate is cut further (players will find the balance that suits them). However, the approach described would give the German player continued hope, while the one he chose left his imposing armed forces up a creek without a paddle.

Before leaving the *Stalingrad* example, we can use it to illustrate the earlier discussion of developments in game design since the 'classics'. Very few modern games would allow a massacre of the attackers at 2–1 (even just once in six attacks), and the Combat Results Table is likely to give a range of partial losses and varying retreats at these moderate odds. The *Stalingrad* map is rather bare for a non-tactical game: all that open terrain was in reality quite varied, and a newer game would probably include roads, minor towns and other terrain modifications. The units are unrealistically uniform in strength, and a designer would now be expected to have done his homework on a more accurate variation of strengths, which would add interest to the game by making the problem of deciding which front should get which units still more acute. This would be reinforced by allowing different types of unit to play different roles: armour would have some kind of breakthrough potential against light infantry defence, but would tend to be particularly susceptible to weather and terrain problems. There is also the question of air forces, and to a lesser extent naval units. Finally, the victory conditions are not very subtle. Shouldn't Moscow have a special value, for instance?

As will be seen, many of the innovations have the effect of making the game more realistic but more complex. Many experienced players find that this is necessary to preserve real excitement and interest: they feel that it is a little *too* easy to be good at a 'classic' or some of SPI's simple Quadrigame series, once one has played a number of games against a competent opponent. This objection does not apply for a new player, who will find plenty to occupy his attention and can have a lot of fun in the process of discovering the key factors of the game, especially if their regular sparring partners are no great experts either. The nice thing about a 'classic' is that it is fast-moving and gives a high ratio of tension and challenge to the time needed to learn the rules. More recent games have a higher potential in many ways, but the newcomer may find that learning the innovations at the same time as the traditional features is slightly overwhelming.

So far, we have also been looking mainly at the AH 'classic' line, because this is a group of games easy for beginners and with broadly similar rules. SPI, however, have their own range of products suitable for newcomers to start on, including an especially cheap game designed specifically for this group of players: *Napoleon at Waterloo*. This is very similar in flavour to the 'classics', and is eminently suitable as an introduction to the hobby, although it has some features which are rarely seen in other games, such as a 'domino effect' retreat rule, under which a retreating unit can displace a whole line of its brethren if they were packed closely behind. The Quadrigames (sets of four games on a theme with similar rules; the individual games can also be bought separately) are also good bargains with simple rules and very smooth play, combined with some challenging problems, though again they have gaps in realism.

Another game popular with new players as well as experienced ones wanting a quick finish is *Winter War*, dealing with the Soviet–Finnish war in 1939–40, in which the Russian units in real life performed so poorly that Hitler was encouraged to believe that the Red Army would stand no chance when the German invasion was launched.

Illustration 6 shows the *Winter War* map, with units set up for the first turn. The Finnish forces start the game inverted, so the Soviet player will not know the defensive strengths when he commits his units, and there are six blank Finnish counters to increase the possibilities of bluffing.

The map shows the situation after the Russians have set up their forces, and the Finns have revealed their dispositions. The Finns have succeeded in luring the inexperienced Russian player into a blunder of monumental proportions. Exactly what the Russians have done wrong is the problem for the reader in this chapter.

To enable you to answer it, here are the facts which you need to know about the game:

1) *Victory conditions:* There are ten turns, during which time the Soviet Union needs to capture objectives worth 61 points to win, and 31 points to draw. Any lower total gives a Finnish victory.
The main objectives are:

Petsamo (far north): 30 points.
Oulu (centre): 30 points.
The whole Mannerheim line (southern fortified strip): 40 points.
Viipuri (town behind the Mannerheim line): 25 points.

2) *Movement:* Rail movement is *free* and unlimited within one's own country, prohibited outside. The crossed lines are rail lines; thus the Russians have one rail line, running from Leningrad to Murmansk, while the Finns cover most of the country except the far north. Minor lakes and rivers cost an extra movement point to cross; mountains and water hexes are prohibited to all units. The Finnish 6-6-2 and the Soviet 20-12-2 units are required to start approximately as shown and must stay in the southern 13 rows. One can have two units per hex in the southern 13 rows (starting with that marked A); only one further north. The various swamps and rivers on the map slow up movement so much that it is a reasonable approximation to say that movement must be along rail lines or along the roads (solid black lines), exceptions being the chain of open white hexes running across the centre to Oulu, and the similar chain leading to the rail line north of Lake Ladoga. Norway and Sweden are impassable. All units stop on entering enemy zones of control, and cannot move through them.
3) *Combat:* Units in the two southern Finnish strips are doubled on defence (the Mannerheim and Ladoga lines), as are units in friendly towns and behind rivers. Soviet units must attack all Finns to whom they are adjacent, while attacks are voluntary for the Finns. If the defending unit is a Finnish 1-1-3, it can retreat one hex before combat would have taken place, so long as it is north of row A; if this happens, then one of the attackers can advance after it (though not attack it) into the vacated hex. This means that the light Finnish units can hold the advance to one hex at a time. If attacks *are* carried out and the result is the elimination of the defender, then again the attacker can advance one hex.

TERRAIN EFFECTS CHART

TERRAIN	EFFECT ON MOVEMENT (cost to enter)	EFFECT ON COMBAT (Defender on given terrain)
Clear	One Movement Point (MP) to enter.	No Effect.
Road	One MP to enter if entered through Road hexside.	No Effect, apply effect of other terrain in hex.
Railroad	See Movement, Case "J"	No effect, apply effect of other terrain in hex.
Finnish City	Apply cost of other terrain in hex.	Finnish/Allied units double their Defense Strength.
Soviet Supply City	Apply cost of other terrain in hex.	Soviet units double their Defense Strength.
Finnish Fortification	One MP to enter	Finnish/Allied units double their Defense Strength.
Finnish Swamp	Finns: one MP to enter. Soviets: two MP to enter.	No Effect.
USSR Swamp	Finns & Soviets: Two MP to enter.	No Effect.
Minor Lake hexside or River hexside	One additional MP to cross.	Defensive Strength doubled if attacked exclusively through Minor Lake/River hexsides.
Minor Lake whole-hexes	Entry Prohibited.	No Effect.
Major Lake or Sea hexsides	Traversing Prohibited	May not attack across. Zones of Control do not extend across.
Ice hex	Two MP to enter. Units may enter only during last three Game-Turns.	No Effect.
Mountains	Entry Prohibited	No Effect.
Norway & Sweden	Entry Prohibited (in Standard Game)	No Effect.
Any hex north of Row "A"	Stacking prohibited (except for Ski units). 20-12-2's and 6-6-2's may not enter.	Apply effect of other terrain.
Row "A" and southward	Units may stack two per hex.	Finns may not retreat before combat.
Finland	Soviet NKVD units may not enter.	NKVD may not attack across Russo-Finnish border.

COMBAT RESULTS TABLE

(1-4 or worse not permitted; if unavoidable, result = Ae; odds greater than 6-1 are treated as 6-1)

die	Combat Odds (Attack Strength — to — Defense Strength)									die
	1-4	1-3	1-2	1-1	2-1	3-1	4-1	5-1	6-1	
1	Ae	Ae	Ae	Ae	Ar	Ne	Ne	Dr (Ne)	De (Ne)	1
2	Ae	Ae	Ae	Ar	Ne	Ne	Dr (Ne)	De (Ex)	De (Ex)	2
3	Ae	Ae	Ar	Ne	Ex	Dr (Ne)	De (Ex)	De (Ex)	De (Ex)	3
4	Ae	Ar	Ar	Ex	Dr (Ne)	De (Ex)	De (Ex)	De (Ex)	De (Ex)	4
5	Ae	Ar	Ne	Dr (Ne)	De (Ex)	De (Ex)	De (Ex)	De (Ex)	De	5
6	Ae	Ne	Ne	Dr (Ne)	De (Ex)	De (Ex)	De (Ex)	De	De	6

(Use results in parentheses when resolving attacks against Finnish units in Fortified hexes. (If a result is printed in parentheses.))

Ae — ATTACKER ELIMINATED: all Attacking units eliminated from the map. Defender may not advance.

Ar — ATTACKER RETREATS: all Attacking units must retreat one hex. Attacker's choice of retreat hex.

Ne — NO EFFECT: nothing happens, units remain where they are.

Dr — DEFENDER RETREATS: all Defending units retreat one hex. Defender's choice of retreat hex. Attacker may not advance.

De — DEFENDER ELIMINATED: all Defending units eliminated from the map. Attacker may advance one of his Attacking units into the hex vacated by the eliminated Defending unit, even if the hex is in another Defending unit's Zone of Control.

Ex — EXCHANGE: Defender removes all units; Attacker removes at least as many Attack Strength Points as the Defender removed in Defense Strength Points (use basic, not doubled strengths). If Attacker has any Attacking units remaining, he may advance one into the hex vacated by the Defending unit, even if in another Defending unit's Zone of Control.

RETREATING: Units forced to retreat into Enemy Zones of Control or in violation of the Stacking limit are eliminated instead. However, a unit can retreat on top of another Friendly unit on or below "row A" even if in an Enemy Zone of Control. Units may not retreat into impassable or prohibited terrain. Advances and retreats are not considered as regular movement and do not expend any Movement Points. Note that the only instance in which the Attacker may advance as a result of combat is when the Defender is eliminated by an Ex or De result or if the Defender has chosen to retreat before combat.

VICTORY POINT SCHEDULE

VICTORY POINTS AWARDED TO THE SOVIET PLAYER

Condition	Point Value
Soviets capture hexes in the row West of Viipuri	+5 for each hex
Soviets capture Viipuri	+25
Soviets capture Petsamo	+30
Soviets capture Oulu	+30
Soviets capture all four hexes of Mannerheim Line	+40
Soviets eliminate or clear out all Finnish units south of Row A inclusive	+50
Soviets eliminate one Allied 3-2-2	+5 per unit

VICTORY POINTS SUBTRACTED FROM SOVIET PLAYER'S SCORE

Condition	Point Value
Finns capture a Soviet Supply City (other than Leningrad)	10
Soviets have less than 7 combat units within 5 hexes of Murmansk at end of their Player Turn, from Game-Turn 4 onward	10 per Soviet Player Turn
Finns capture Murmansk	20
Finns capture Leningrad	100

Note: see "capture" definition.

COMBAT RESULTS TABLE

die	1-4	1-3	1-2	1-1	2-1	3-1	4-1	5-1	6-1	die
1	Ae	Ae	Ae	Ae	Ar	Ne	Ne	Dr (Ne)	De (Ne)	1
2	Ae	Ae	Ae	Ar	Ne	Ne	Dr (Ne)	De (Ex)	De (Ex)	2
3	Ae	Ae	Ar	Ne	Ex	Dr (Ne)	De (Ex)	De (Ex)	De (Ex)	3
4	Ae	Ar	Ar	Ex	Dr (Ne)	De (Ex)	De (Ex)	De (Ex)	De (Ex)	4
5	Ae	Ar	Ne	Dr (Ne)	De (Ex)	De (Ex)	De (Ex)	De (Ex)	De	5
6	Ae	Ne	Ne	Dr (Ne)	De (Ex)	De (Ex)	De (Ex)	De	De	6

FINNISH INITIAL FORCES

x5 blanks

4-4-2 x1 — May be deployed anywhere in Finland

2-2-3 x3, 1-1-3 x10

6-6-2 x7 — Must be deployed on Fortification hexes

FINNISH REINFORCEMENTS

The horizontal track indicates the Game-Turn in which the reinforcements are to be brought on. Use a coin as a marker to indicate which Game-Turn is being played. The pictures of the units under a given Game-Turn, indicate the types of units to be brought on. The multiplier number under each unit picture indicates the quantity received of that type of unit. Players may deliberately delay the appearance of some or all of the units which they are to receive in a given Game-Turn.

Finnish reinforcements enter the game during the Finnish Reinforcement Phase at any Finnish city which is controlled by the Finnish Player, i.e., occupied by Finnish units or through which Finnish units were the last to pass. Finnish units enter the game in an unstacked state. The act of placing the units on the map does not expend any Movement Points. Reinforcement units may exceed the stacking limits when initially placed, but they must conform to the stacking limits by the end of the Movement Phase.

Die Roll	Scenario Table (roll the die once ...			
1,2,3	•	•	•	•
4	•	C	A	A
5	•	D	B	A
6	D	E	C	B

Game-Turn	1 (30 Nov 1939 – 9 Dec)	2 (10 Dec – 19 Dec)	3 (20 Dec – 29 Dec)	4 (30 Dec – 9 Jan 1940)
REINFORCEMENTS		6-6-2 x1	2-2-3 x1	
	0-0-3 x5, 1-1-3 x2	2-2-3 x2	1-1-3 x2	

6 Mapboard of *Winter War*. The position is post-setup, pre-invasion.

4) *Reinforcements:* Moderately large reinforcements appear on both sides during the game: the Russians get a 20-12-2 on turns 3 and 4, for instance, while the Finns get a 6-6-2 on turn 2, to be used south of Row A, a 4-4-2 on turn 6, and three 2-2-3s on turns 2–3.
5) *Combat results table:* A 1–1 attack is very dangerous, as it can lead to elimination of the attacker. A 2–1 is safe (though the attacker may have to retreat) and the minimum to affect fortified hexes.

There are also supply rules, which we will omit here to avoid excessive complication (the Soviet supply units included in the game are not shown on the map for the same reason) and some other minor points not affecting the strategic considerations. Soviet border defence units have been omitted.

Problem: Napoleon Clausewitz Smith, the Soviet player, explains his dispositions to you like this: 'Well, ah, um, I saw all the units face down, of course, I mean I couldn't see them, you see? OK. The Finns have six blank counters, and I worked out where it would leave the biggest hole if the counter turned out blank. That was in the centre, near Oulu, where he's only got one unit every few hexes, right? Then up north he's just left a great big gap, which I'll be marching into this turn with a couple of 6-4-2s. It turns out those counters in the middle weren't blanks, but he's so weak there my 6-4-2s will just chew him up, or push him right back if he retreats before combat, the rotten coward. Far north I've got a big force around Petsamo and can reinforce it from the centre if necessary, and in the south I've got the 20-12-2s, which should be able to take two or three hexes of the Mannerheim line during the game. That's Oulu, Petsamo, most of Mannerheim at 10 points per hex, 80–90 points, and 60–70 even if I get no luck in the dice and miss one. Can't lose, man!'

Find all the things wrong with this penetrating analysis. (See Appendix A for answers.)

4 RESERVES AND BUILD-UP

Who needs reserves? Many players act as if the concept had never been invented. 'Here's a spare 3-3-6, darn it! Must be something I can do with it. Oh, let's put it in the big battle. Can't do any harm.'

Infuriatingly, this works perfectly well in some games, at least some of the time. In *Stalingrad*'s first turns, there is something wrong with an offensive which doesn't use virtually every German unit to get optimal odds. The Russians are not going to get through the lines on the next turn, and nearly every section of the central front can be reached in one turn by using the rail line. Later on this ceases to be true, as we saw in Chapter 3, and it sometimes becomes necessary to keep some forces hovering in the middle between the fronts, ready to dash either way as the situation requires. In general terms, however, the armies must be kept busy in *Stalingrad.*

When a *Stalingrad* player reared in this tradition progresses to *Third Reich* and tries to play in the same way, he will lose the game in a single turn. I speak from bitter experience. I played my first postal game of *Third Reich* after a decade of wargames experience on every type of design and ... well, I'll show you what happened.

The game deals with the whole of the Second World War in Europe at corps level. There are up to six players: Germany, Britain, France, United States, USSR and Italy; when minor countries are attacked, their defences are conducted by the nearest player hostile to the invader. There are air and naval units at a rather abstract level; to simplify matters, I will leave these out of the discussion, as they cancelled each other out on the occasion in question.

The nasty thing about defending in the game is the armoured breakthrough rule. You may have a powerful defence line stretching shoulder to shoulder across the front, but it can be virtually annihilated like this: the enemy attacks your weakest unit and destroys it.

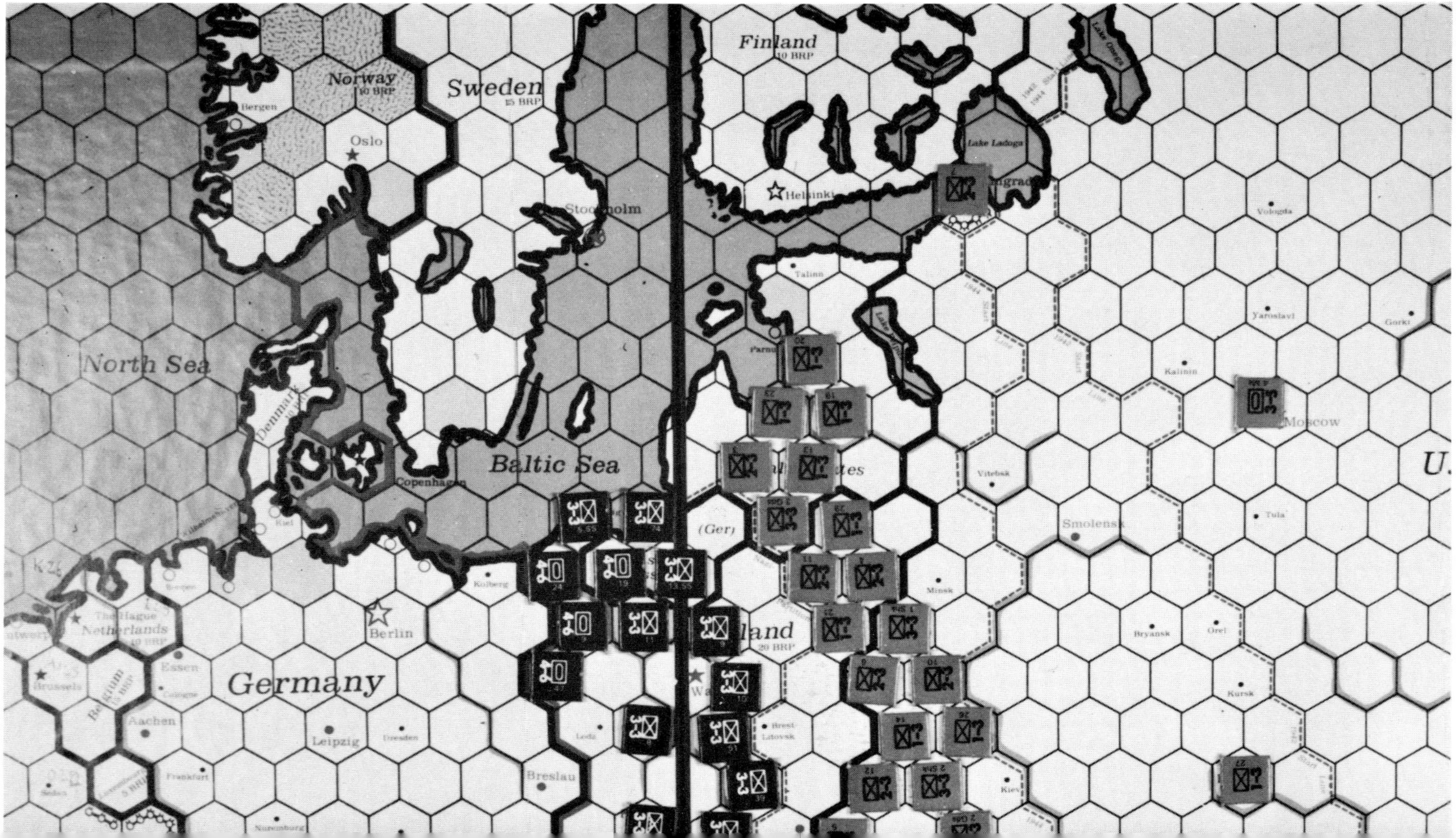

Finland
10 BRP
Norway
10 BRP
Sweden
15 BRP
Bergen
Oslo
Stockholm
Helsinki
Lake Ladoga
Lake Onega
Talinn
Vologda
Yaroslavl
Gorki
Kalinin
Moscow
North Sea
Denmark
Copenhagen
Baltic Sea
Kiel
Vitebsk
Smolensk
Tula
(Ger)
Kolberg
Minsk
The Hague
Netherlands
10 BRP
Berlin
20 BRP
Bryansk
Orel
Essen
Brussels
Cologne
Germany
Kursk
Aachen
Leipzig
Dresden
Lodz
Brest-Litovsk
Frankfurt
Breslau
Kiev
Nuremburg

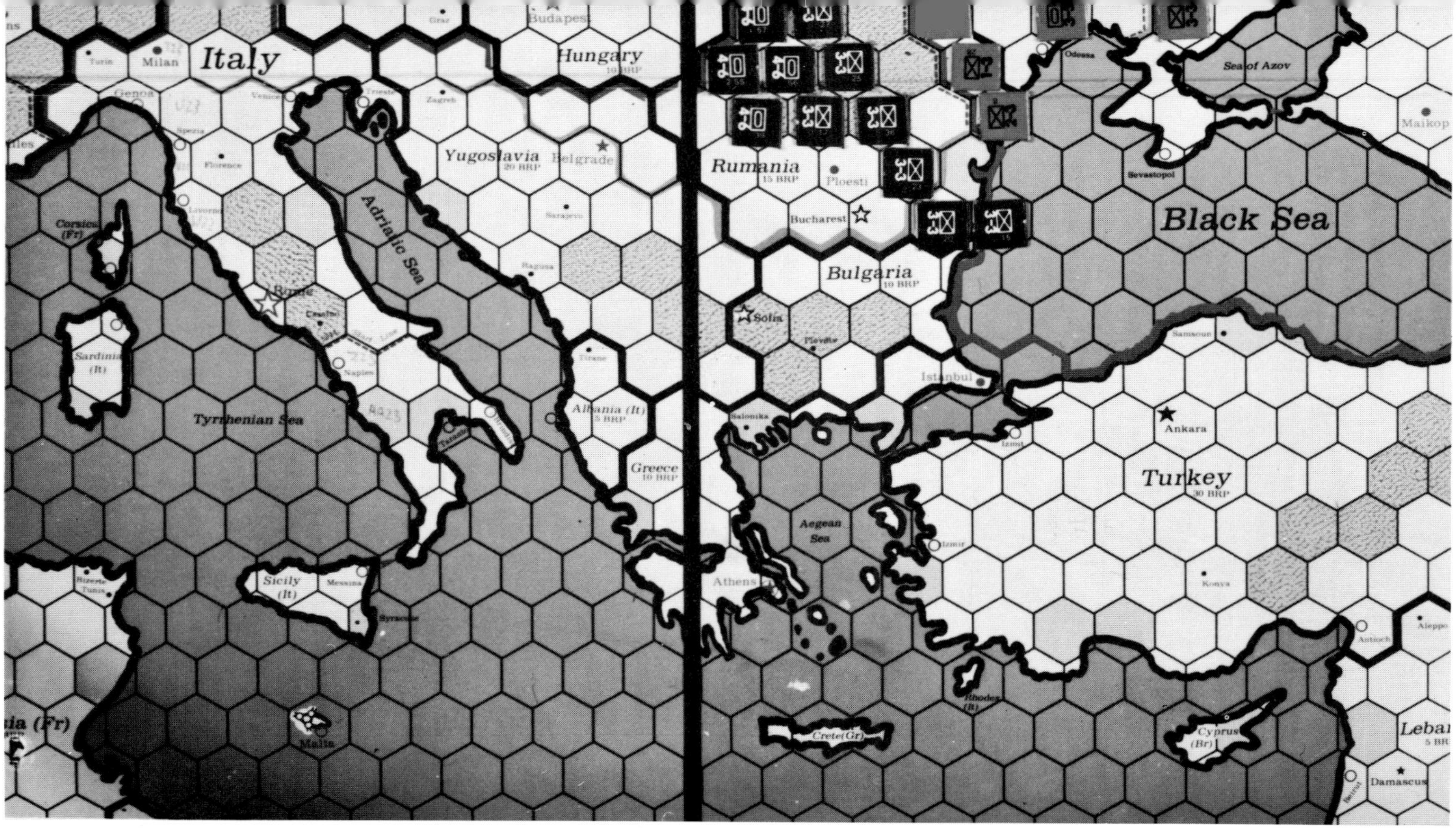

7 Detail from *Third Reich.*

Armour associated with the attack can be placed on the dead body of your unit, without any stacking limit, as they aren't staying there. Now they are given another full move, the 'exploitation phase', and use this to fan out up and down your defence line, from behind. Next turn you find that all of your units in the line are cut off from their sources of supply. Unsupplied units cannot move, and die at the end of the turn. Failing some gallant rescue action by troops outside the pocket, you have lost 25% or so of your army. This is assuming that you remembered to garrison the key centres behind the line; many careless Frenchmen have lost Paris in 1939 by neglecting to see every possible breakthrough. If the capital is lightly defended, this may be supported by a paratroop corps dropped in the area.

The only answer is to have a double line of defenders. Then, the armour exploiting a victory in the first row will face a second defensive wall. He can't exploit twice, and you're safe.

Well, I *knew* that. The hints in the rules point it out. I would no more try to defend in *Third Reich* with a single defence line than I would try to fly to the moon. I had the USSR, and had been arming to the teeth since 1939, while a brilliant German–Italian combine steamrollered France, the neutrals and England in rapid succession, taking advantage of a series of slight blunders. Finally, the megalith massed threateningly on my frontier, and I knew it was my turn. Illustration 7 shows my defence lines. Some other Soviet units were engaged in defending Turkey (conquered earlier) and in garrisoning certain key towns. The line was pulled back slightly to neutralize some of his air units behind the front.

All units are doubled on defence in the game, so e.g. a 1–3 is what in *Stalingrad* would be called a 1–2–3. The double column of defenders in the row opposite Warsaw and the two lower rows are in the Pripyat marshes, and tripled instead of doubled (there is no movement delay). Only armour has a zone of control in this game, and enemy forces can move through and out of it at the cost of three movement points; this is why I put my two armoured units two hexes away from the front. Suppose the Germans massacre the 1–3 next to the southern end of the front, and exploit from there with their 4-6 armour. One movement point gets them into contact with one of my armoured corps. Three more enable them to squeeze into the gap between my armour, by moving through my ZOC. But that leaves only two movement points, which is not enough to progress further through my ZOCs. Had my armour been one hex nearer the front, the Germans could have moved straight into the gap (movement cost three) and out behind my southernmost tank corps, 13Me (three more); the German ZOC would then have reached to the Black Sea, cutting off the supply of 13Me.

Problem: The German forces which accomplished my downfall are among those shown on the map. You can assume for the purposes of the problem that a 1–1 gives a good chance of eliminating the defender (and possibly the attacker too – a 2–1 greatly reduces this risk). The Soviet player can survive heavy losses, as he can use saved-up economic resources to replace losses quickly. But if most of the army is cut off from supply, that would be a different matter! You have eight 4-6 armour units, one paratroop 3-3, and as many 3-3 infantry as you can use. The paras can be dropped anywhere within six hexes, but if it drops on any enemy unit it must attack it. Armour can exploit from a victory if it was adjacent to the original attack force; one just moves it (at no cost) to the victory hex and exploits from there. Apart from this, the stacking limit is two units. ZOCs extend through enemy units.

Although this is really a tactical problem, you should be able to see what the German player (*Third Reich* expert Michael Hardwick) did. If you can, you'd have beaten me too (I have adjusted the situation slightly for easier presentation). What had I forgotten? The answer is in the Appendix.

In *Third Reich*, therefore, one may need either a triple line, which is normally impossible, or a second line sufficiently far back from the first to avoid German air cover being available. This is rather unusual, but there are many games where armoured units are allowed a second movement phase after combat, so that the attacker can hope to blast the obstacles out of his way and break through the enemy lines for the second phase. A double line is the only effective defence against this.

A different kind of double-line formation is useful in many pre-twentieth-century games, such as *Nordlingen*, discussed in Part V in more detail. The point here is that most games of this type have some system of attrition of front-line troops, and the units have to be pulled back from the front to recover their full strength. This is also true of nearly all games with casualties by step-reduction, where unit counters taking casualties are replaced by new counters for the unit with lower combat factors, and can be restored by the addition of replacement units. To give the recuperating units some peace and quiet while they recover, it is preferable to have a second row of 'healthy' units to take their places in the front. If you maintain gradually deteriorating front-line forces indefinitely, the enemy attacks will be at progressively better odds, accelerating your rate of losses.

Holding a pool of all-purpose reserves back from the front is very tricky and depends altogether on the position. A reserve should be maintained under the following circumstances:

1) The enemy may break through at several different points on his next turn, and he will be able to choose which in the light of your move, or his choice is unpredictable for some other reason.
2) It is not possible to be sure that units committed to the front will be able to cover all the reasonably likely contingencies.

A typical case is when there is a road or rail centre behind the lines, leading to different sectors. Movement from sector to sector across country may be difficult, and going back to the junction and then setting off again for another sector may be too slow; Brest-Litovsk in *Stalingrad* has very much this sort of position in the later stages of the game, as the midpoint between the fronts. If several sectors are threatened, then it may be best to keep a sufficient force back to repulse penetration anywhere. As this reserve force constitutes a threat to the probably important troops which the enemy will need for his alternative breakthroughs, it may actually compel him to postpone all the assaults. To have three weak sectors and an effective counter-punch to penetration of any of them is infinitely better than shoring up two of the sectors to impregnability, and leaving weaknesses in the third, since your opponent will, of course, choose to attack this one.

The converse applies for attack. Placing a group with a strong total attack factor at a central communications junction poses a threat to every part of the enemy line which is in reach. This tactic is best suited to games where the Combat Results Table (CRT) is 'bloodless' and dominated by retreats for each side. In these games, the decisive advantage tactically is gained by positional manœuvre, especially by breakthroughs leading to the surrounding of enemy pockets. This means that it is worth passing up some attacks at fairly good odds if, by holding back a group of units, you can hope to tear a hole in the other player's lines on the next turn. If the CRT is bloodier, with heavy losses and even whole units being eliminated at a stroke, then you will usually want to pack nearly all your units into the front; the 'classic' CRT is one of the bloodiest of all, which is why *Stalingrad* players are reluctant to keep large reserves.

Another reason for holding back is to wait for reinforcements which will improve the chances of the assault. It is essential here to keep a careful eye on strategic timing. If the victory conditions look feasible with a little time in hand, then it is worth waiting for improved odds if the wait is allowed for in the timetable, but the further you are behind schedule, the more prepared you should be to take risks with a weaker attack. There are no prizes for *nearly* winning Waterloo; Napoleon did ('a damned close-run thing') but much good it did him. Always take the best chance of victory which remains possible.

Seelöwe is SPI's game on the hypothetical German invasion of Britain in 1940 (another simulation of this is contained in GDW's *Their Finest Hour*). Illustration 8 shows the main battle situation in a game of the July (East Coast) Scenario, which assumes that the Germans had anticipated the rapid fall of France and were ready to follow up with a fast invasion of Britain, with the defenders unprepared. The position is half-way through the third German turn (out of fifteen). The initial turns have been bloodless, with an almost unimpeded German build-up on the eastern coast between Yarmouth and Southend, though rough seas forced the most recent wave of reinforcements into ports between Lowestoft and Harwich, inconveniently for the drive on London. The German objective is the capture of as much as possible of London, while retaining the captured ports to the east.

The British have been building up strength in London, rather than trying to confront the invaders in the flat, open country to the north-east of the capital. There are three reasons for this:

1) The units on the map which are inverted (white) have not yet succeeded in mobilizing, though they can be expected to do so within the next turn or two. Until these units, and others being rushed in by train from various parts of the map, are available, the British forces would risk a savage mauling.
2) The game assumes that the Luftwaffe have attained air superiority, and most of the RAF have been pulled north; after some losses to anti-aircraft in the first turns, there are now eight Luftwaffe ground support air units available to three for the RAF, though the RAF will get eight more later in the game as the full force is flung into the battle for London. Air attacks halve combat factors and reduce the movement factor to one if they are successful in disrupting the defenders (which five attackers in this part of the map will be, five sixths of the time). While disruption is not permanent, it is crippling while it lasts. But air attacks are ineffective against units in towns, which is why not only the London defence but nearly all the reinforcement units waiting to be sent to the front are in towns. This turn, a 2-3 which had been in Chatham (south-east of London) and moved into the open to cover the Thames crossing immediately east of London was disrupted (signified by the

September Deployment (Scenarios N and H)
British Initial Forces (Order of Battle)
July Deployment (Scenario J)
13-8
10-8
8-8
8-3
6-5
6-3
3-6
3-6
1-2
5-4
6-4
1-3
(1)-2
(1)-2
(1)-2
4-5
5-4

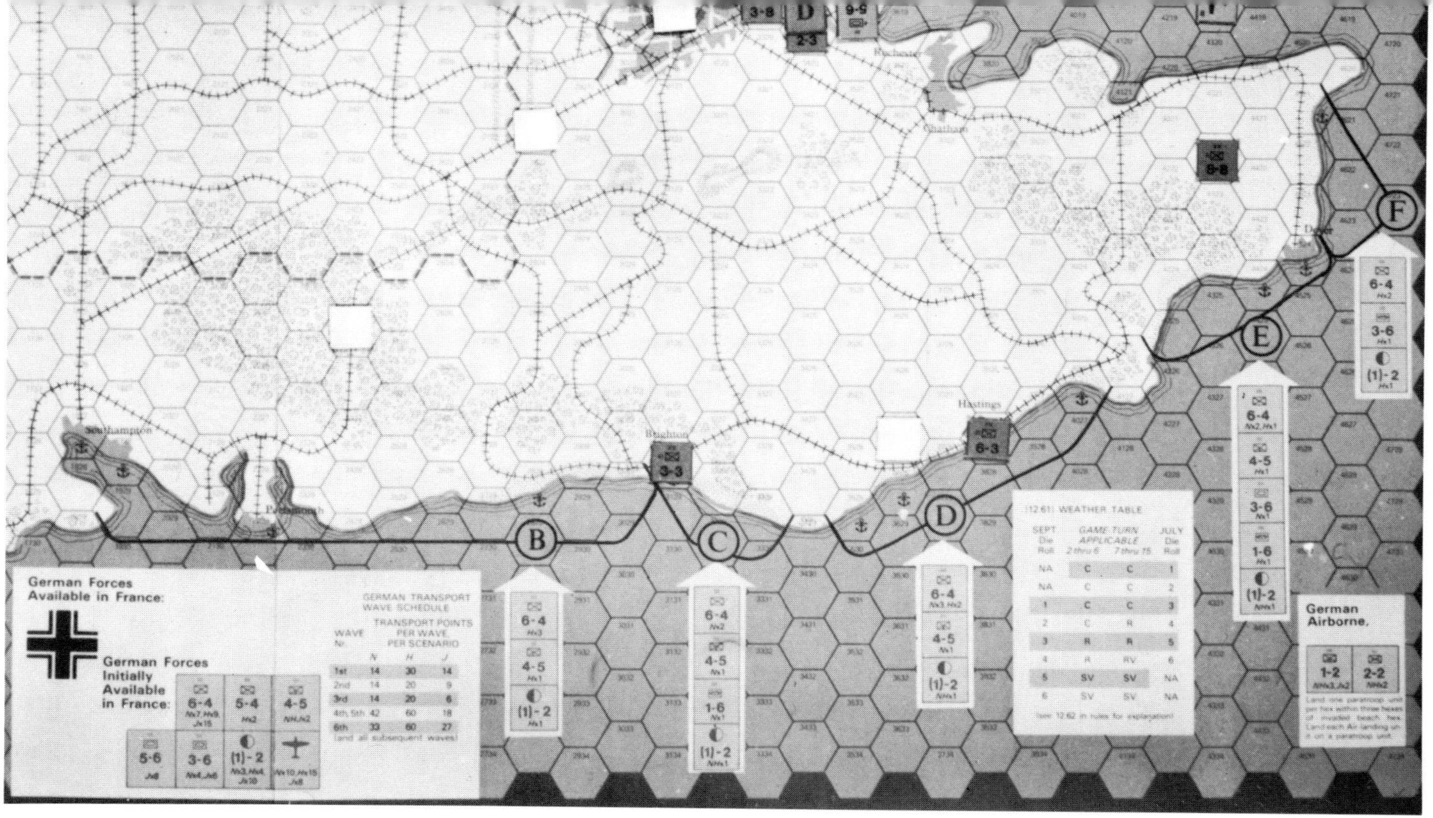

8 East half of mapboard of *Seelöwe.*

D counter), but a 4-unit air attack on the 8-8 further east (which had been disrupted earlier) failed, and this can now reach Chatham if the Germans don't get there first.

3) The CRT is fairly bloodless, with retreats dominating, and has the important addition that units in towns do not have to follow retreat results. Despite their numerical inferiority, the British can expect to have a good chance of holding their own in London. A 3–1 is needed to get an even chance of eliminating the defenders in a battle, and this will commonly be in an exchange, losing an equal number of attacking factors. A 6–1 is required for an even chance of pure elimination.

It should be noted that surrounded units are usually halved for lack of supply (unless they are British forces in towns), while their attackers are doubled. In general, the British have few supply problems, while the Germans require a (1)-2 supply unit within five hexes to attack, and then expend the unit (new supply can be brought ashore, but its movement factor of 2 makes its advance slow).

Entrained units need a stationary turn to disembark, so the reinforcements are not going to be much good, except possibly for defence, for another two turns. However, there are 27 factors now on the way, so the British situation will be much easier if they can hold on that long.

Armoured/motorized forces move after combat. So far, the two German 5-6 armoured units have held back, waiting for the slower infantry to mass in front of London, but this turn they strike at the weak points of the defences. Attacking is always voluntary in *Seelöwe*.

In North London, the German 5-6 and 4-5 attack the 2-5 at 4–1. It would have been useful to have the second 4-5 further east in the assault, making the odds 6–1, but it was seven hexes from the defending unit, so could only reach two hexes away. As it stands, the attack has a one-third chance of eliminating the defence, in which case the 4-5 will advance into its vacated hex. The nearest 3-6 will move to fill the gap north-east of the captured position in the second movement phase, while the 5-6 will push one hex south into London. The fast British 1-12 can still loop round the back of the German line to surround the 4-5, but in this game, supply can be traced over friendly units through enemy ZOCs, so the 3-6 will enable the 4-5 in North London to stay supplied, and the risk of taking the 1-12 out of London would be considerable, though probably necessary. The attack also has a one-third chance of exchange, in which case the 4-5 will be taken off and the 5-6 (very nervously) will advance. Either way, one hex of Greater London will have been captured, with the defensive advantages as well as the prospective victory points accruing accordingly.

Meanwhile, to the east of London, the second 5-6 and a 5-4 assault the disrupted 2-3 on the south side of the Thames, at 7–1. Two pips are taken off the die roll for the defensive advantage of the river, but it is certain that the 2-3 will be forced back one hex. The German will push it south-west, followed by the German 5-4. In the second movement phase, the armoured 5-6 crosses the Thames and stops. It is extremely tempting to go into Chatham, denying it to the British 8-8 and giving the Luftwaffe another chance at it. However, the 8-8 and a 3-8 from London could then surround the 5-6 and pound it at 8–1 (with the surrounded effect in full force), and not even the town advantage would save it. This seems an excessive price. The 8-8 cannot slip behind the unit if it stays just south of the Thames, as there is an extra movement cost of one point for moving into an enemy ZOC, so 9 movement factors would be needed.

The example shows the limitations of the second movement phase as well as its advantages. On their own, the 5-6s cannot stick their necks out too far, or they will be isolated and destroyed by the British reserves kept back for this very purpose. As the new armour reaches the front, the Germans will be able to push further on the second phase (and more reinforcements can be expected on each turn, though the landing places will be increasingly far from the action), but by then, the British forces will have rigged up a better defence line. So long as reserves are available which are sufficient to mop up any breakthrough, the defence should probably hold, since the Germans are severely handicapped by the supply problem (on this turn the most advanced (1)-2 supply unit will be removed to fuel the two attacks) and the ineffectiveness of the Luftwaffe against troops in London.

In passing, it may be remarked that there are two other *Seelöwe* scenarios, set in September 1940 (with correspondingly rougher weather, interfering with air support as well), with landings along the south coast. Chatham again plays an important role, as a staging area where British forces can build up before striking at the crucial beach area around Hastings. Shades of William the Conqueror.

Before leaving the subject of reserves, it is worth mentioning the special case of hidden-movement and simultaneous-movement games. As emphasized earlier, one needs reserves when one is uncertain what the other player is going to do next. If you are not even sure where he is at the moment, this applies to a particularly marked degree, and if movement is simultaneous, then you have to work out what he might be able to do to you in *two* turns – the one he is making at the same time as you, and the one he makes in the light of what you've done. True, you will be moving again then too, but if you have sent your entire force in the wrong direction this turn, it may be too late to repair the damage. *The greater the uncertainty, the more reserves you need.*

5 POLITICAL AND ECONOMIC AFFAIRS

If we are to believe some of the historical accounts, war used to be delightfully simple for the noble antagonists. First they would raise a force of local peasants, offering a handsome uniform or the prospect of goodly spoils, and reminding them that loyalty to their kindly and illustrious liege was the highest virtue in civilized society. Then they would march over to the rival's stronghold and fix an appropriate time for the battle.

'Nine o'clock suit you?'

'I've been having some late nights recently. Make it ten.'

Well, perhaps it wasn't quite like that (though Cromwell won a celebrated victory partly by attacking before his opponent expected the battle to start). But it seems certain that some of the great generals of old would have quailed before the sheer complexity of modern warfare between nations. How much effort should be put into destroying the enemy ball-bearing industry? Is the merchant fleet sufficient to run a submarine blockade? Should heavy industry be bodily moved a thousand miles east to escape capture? Has the enemy fully mastered the use of his new weapons?

More familiar might have been the political dilemmas. Is Italy going to come into the war or not? Why is my neighbour fortifying his frontier with me – because he wants a fall-back position for an invasion of my territory, or because someone has told him I have designs on him? And if the latter, can it be my other neighbour who told him, and if so, should I revise my cooperation with him?

Wargames need to simulate every aspect of conflict, so all these factors appear in different games, and it seems reasonable to predict that the trend over the next few years will be to more and more integration of political and economic factors into the games.

POLITICS

The trailblazer in the political field was *Diplomacy*, a seven-player game which is not a wargame at all except in the broadest sense: the military/naval rules are very elementary indeed, with just 'armies' and 'fleets' which move around big provinces on a map of Europe in 1900, the fortunes of war depending on possession of supply centres in each country.

The charm of *Diplomacy* is the player interaction, which reaches preposterous depths in fiendish deviousness and cunning, as each of the players attempts to bend the other six to his nefarious plans for continental domination. An anecdote from a postal game will serve to illustrate, though a book could be written about this game alone. I had drawn Italy, and heard that my eastern neighbour, Austria-Hungary, was involved in delicate negotiations with Russia and Turkey, in which trust on the next move would be vital all round. I wanted to attack Austria with Turkish help, and Austria knew it, having had photocopies of incriminating letters from me passed on to him by third parties. The problem was to (*a*) get Turkey's support; (*b*) catch Austria off-guard against Turkey; (*c*) make Austria misguess my offensive plans.

I wrote to Turkey, with three separate letters. Letter A asked for his support against Austria, and declared my intention of attacking Vienna. Letter B asked for his support, declared that I would attack Trieste and *not* Vienna, and asked him to send letter A to Austria to gain his trust and make him believe that I was attacking Vienna. Letter C asked for his support, announced, truthfully, that I was attacking Vienna, and asked him to send *both* A and B to Austria, with a covering note promising support against my perfidious plans; it seemed certain that Austria would not think of both letters being faked, and would trust Turkey, as well as covering Trieste rather than Vienna. I would then get Vienna and Turkey's back-stab would succeed.

The result was typical of the subtle flavour of the game: Austria believed Turkey, and swallowed the story, without suspecting the existence of letter C. *And yet* – something, somewhere, was in the wind this turn. He sensed it, without knowing exactly where the pitfall lay. Austria was an old hand, and trusted his instincts. He threw his whole force on the defensive, and covered both Trieste and Vienna, saving his homeland. He survived my attack, and Turkey's attack, and went on to prosper, while I was eliminated by a ferocious French assault. *Sapristi!*

The first multi-player wargames were not a great success. *Strategy*

I simulates war through the ages, with different rules but the same abstract map for each period, and up to eight players with counters which may mean tanks or cavalry, depending on the era. An ambitious game with a magnificent range of rules for the numerous situations covered, the player interaction rarely reaches the extent of *Diplomacy*, perhaps because the rival forces are a little too abstract to catch the imagination and inspire the players to special efforts. *Origins of World War II* suffers from other problems: it is heavily unbalanced in the historical simulation of pre-war politicking, with (oddly) the USA hopelessly placed, and the basic game system of resolving struggles for political influence by battles between 'political factors' on a CRT is probably too similar to military combat rules in other games to carry conviction in the political context. Trying to persuade Italy to ally with France should have a different 'feel' from a combat simulation.

Undeterred, the companies have since produced much better marriages of wargames with politics. *Kingmaker*'s military rules are almost as odd as those of *Diplomacy*, but it generates a lot of entertainment around the table. *Russian Civil War* is another 'fun' game, with practically everyone in sight plotting against everyone else, and *Conquistador*, *After the Holocaust* and *The Plot to Assassinate Hitler* are all multi-player games of different types which have appeared within three months of each other.

More than any other aspect of wargaming, the conduct of political relations with other players has certain fixed principles from which it rarely pays to deviate:

1) Always tell the truth unless you have some good reason not to do so. Some lies are exposed by accident or design of other players, and if you are seen to have lied without purpose, you will not only ruin your credibility with that player but make the others feel you're a pretty hopeless case as an ally as well. It can sometimes even pay to admit that you are ending an alliance, if the other player is not in a position to do much to change his defences anyway. One day you may want to patch things up, and then your earlier frankness will help you carry conviction.
2) Establish relations with everyone, not just the players who can be useful allies at once. There is usually some slight way in which you can help each other early on, and this will give you a very substantial advantage later on when you are competing for influence.
3) Decide if you want to have fun or to win. Too many imaginative ideas are bad for your health, as the other players will think you are trying to sneak something by them even if you are being straight-

forward for a change. Any form of flamboyance is dangerous, and the long-term winners are players who go around with a worried frown, telling people that their situation is very difficult, and things haven't been going at all as they expected; on winning, they attribute it to their loyal allies, good luck, and minor errors on the part of other players. The reason for all this is that attracting attention in multi-player games makes the others gang up on you. On the other hand, perhaps it is more interesting to live colourfully and go down gloriously? It is just possible to have the best of both worlds; Richard Sharp, probably Britain's foremost authority on *Diplomacy*, with such a strong personality that disguise as a baffled rabbit would be impossible, nevertheless succeeds in persuading experienced players that his near-winning positions are actually on the point of collapse, and – the important bit – that such a collapse would benefit everyone but the person currently addressed, and he must do something to help *at once*. With six allies all trying to save you from each other, you prosper with amazing speed.

4) If the game is any good, there will be several reasonable alternative alliances for you. You may have personal preferences for some tactical or strategic reason, but don't let this dominate your negotiations. In a multi-player game, a strong alliance will beat good tactics nearly every time.

5) The ideal partner is on the other side of a lucrative target, and with slightly less good prospects than you. Without stabbing him during your cooperation, which would be counter-productive, there is no need to strive officiously to smooth out all his problems. You want the problems still there when he starts to think about stabbing *you*.

6) Always be willing to offer players in difficulty a helping hand to a respectable place like second or third, and keep such agreements with fanatical fervour; it costs you nothing to ensure that your ally comes second, and if you get a reliable reputation in this respect you can build up an army of small allies who will sweep all before them.

7) Don't bear grudges. If you take things that seriously, what are you doing playing games?

Problem: Illustration 9 shows the map of *Diplomacy*, with the seven main powers, England, France, Germany, Italy, Austria-Hungary, Russia, and Turkey, as well as various defenceless neutrals over whom the big powers squabble. You have Germany at the start of a game. Without going into tactical details, you need to know that Germany is usually involved in a three-cornered struggle with England and France at the start, with possible fencing with Russia over Scandinavia.

9 Diplomacy board, pre-spring 1901, with units in starting positions.

In the south, Austria, Turkey and Russia form another trio, with the Balkans as the prize. Italy may attack France, Austria or Turkey.

You receive the following communications (summarized):

From England: Dear Kaiser, I have been playing for four years, and strongly believe that the Anglo-German alliance is much the best on the board. Moreover, I have been corresponding with the French player, and I'm afraid he is clearly going to attack me. This gives you an excellent opportunity to take him by surprise, which I hope you will take. In return, I will support you in the north.

From France: Dear Kaiser, I am new to this game, and feeling slightly bewildered at present! The only player I have heard from so far is England, who rang me up yesterday. A plausible guy, but he seemed to want me to do all the work. I would be willing to ally with you against him, so long as we share the attack and the spoils.

From Russia: Dear Kaiser, Can you tell me anything about the southern situation? I will in return let you have any information I get about the west. I shall definitely be concentrating in the south – where, I'm not yet sure. Are you willing to cooperate in Scandinavia?

From Italy: Nothing.

From Austria: Dear Kaiser, I know it's a bit unusual, but would you like to join in an early assault on Russia? I think I can get Turkey to join in.

From Turkey: Nothing.

What line do you take to each player, and do you move against England, France, and/or Russia? Who is telling the truth, partially at least, and who is clearly lying in his teeth?

All countries start more or less equally strong, and there is little to choose between the tactical alternatives, though an early attack on Russia needs excellent relations with England and France while they fight each other.

ECONOMICS

There is something about the inclusion of political rules in games which encourages a light-hearted approach. Probably it is the interplay of personalities; it is hard to keep a straight face while your oldest friends vie with each other to assure you of their undying alliance with you in the game, while their eyes flicker over the board, trying to gauge the chance of swallowing you up at a gulp. Economic rules, on the other hand, add a whole new dimension of skill and challenge to the game, and there is a good case for saying that no complex game of twentieth-century warfare is complete without them.

The economic structure is almost always tied to military production, since that is what will affect the fighting. I should very much like to see more games where there is a choice between economic and military priorities, with popular pressure making too much emphasis on armaments difficult, and a chance of winning by low arms spending and clever diplomacy keeping the country out of trouble (and a corresponding military disaster if a war with the neighbouring country does start), but this is not usually possible, with the single exception of *After the Holocaust*. What many games do have is an 'investment' option, enabling you to build up your production apparatus for future military use, rather than roll out the tanks now. Another possibility designers should explore in this context is giving an edge to the most recent production: it is widely believed that the Battle of Britain ended as it did partly because the latest aircraft designs were coming off the assembly lines at exactly the right moment, whereas the German production machine had started earlier and now had slightly older designs. Similarly, Soviet arms production went into top gear some months before the German assault, with the results appearing up to a year later, just in time to save the country from defeat, despite the loss of the European industrial centres.

Third Reich is typical of games with production and investment. Each country starts with a fixed number of 'basic resource points' (BRPs), irreverently called 'burps' by most players. These can be spent on equipping new units, with a naturally higher cost for specialized and technologically advanced units than for infantry; there is also a BRP cost for declaring war, presumably reflecting the cost of mobilization and transferring the economy to a war footing. BRPs not used during a given year are invested, with each nation's growth rate determining the rate of return; the United States naturally has the highest growth rate. BRPs can be gained by occupation of other countries, or by receiving loans or gifts from allies. The total economic structure thus created is subtle and flexible, allowing for a wide range of possible policies.

A simpler approach to production is seen in games like *World War III*, in which the flow of resources is fixed and cannot be varied by investment, but the players have a choice of end-products. Not unnaturally, these take different periods to produce. The table of costs in the game looks like this:

UNIT TYPE	COST	TIME
Fleet with aircraft	6	10
Fleet without aircraft	4	7
Nuclear submarine fleet	4	7

Anti-submarine air group	3	3
Coastal defence force	3	4
Amphibious assault ships	4	4
Merchant shipping	2	3
Land force, 1 combat point	4	3
Supply or port	1	1
Industrial centre	20	8

As is usual in this game, I have omitted nuclear missiles, since the use of these by either side tends to end the game (and incidentally civilization). The game has twenty turns, and resources become available each turn, depending on the number of industrial hexes available to each player, and the current production multiple; at the start, one point is available per industrial hex, but this rises gradually to four by turn 13.

Western Europe (with the curious exception of Norway) is usually quickly overrun by Soviet forces, leaving North America and Japan with 14 industrial centres, so up to 56 points a turn. Here is a typical calculation by the US player on turn 8, when the multiple is 2, so he has 28 points available:

'I seem to be wearing him down on the oceans satisfactorily, so I can forget about naval builds, and my merchant fleet has been rebuilt on earlier turns. Victory conditions require me to regain part of Europe, or some of the Soviet industry elsewhere; I shall need amphibious assault craft and land forces for that. I may need ports or supply counters, depending on where I invade. I need one amphibious unit for each land strength point.

I already have a few amphibian and land strength units. The first thing to build up is amphibians, because they take longest. I reckon I shall need to start invading by turn 14 to get to the targets in time, so I have seven construction phases, two with multiple two, three with multiple three, and two with multiple four. The best programme seems to be this:

Turn 8:	28 points.	Build 7 amphibians, ready turn 12 (for use turn 13).
Turn 9:	28 points.	Build 5 amphibians, ready turn 13 (for use turn 14).
		Build 2 land strength points, ready turn 12.
Turn 10:	42 points.	Build 10 land strength points, ready turn 13.
		Build 1 port, 1 supply unit, ready turn 11.

On later turns, concentrate on land reinforcements, since the amphibious forces can be used again, though some further amphibians should be built to transport larger forces, or for when one wave of amphibians is being transported back to the USA. This schedule means that a maximum twelve points can be ferried in on turn 14. Note that if the player had decided he could wait till turn 15, then his builds would be different right back on turn 10, when he could fit in five more amphibians and have that many more invaders on the day. If he has played his turn 10 already, then he has effectively closed his option to launch a bigger invasion a turn later. The die is cast.

This simple example shows the powerful effect of production rules on strategy: it is essential to gear production to the overall plan, and to do so a number of turns earlier than would be necessary in a game with fixed reinforcement (or instant production) rules. In this case, it is easy enough, as only two main types of unit are involved. But what if the player had also needed an anti-submarine group, or a fleet without aircraft, or a nuclear submarine fleet, depending on the circumstances at the time? What if the composition of the force depended on what forces the Soviet player built in the meantime? *World War III* is not an ideal game as an example here, since there are good reasons why each side will keep their builds in a narrow range of choices which will not be likely to surprise the other player much; *Strategy I*, for instance, offers a better chance of concealing your plans (and even your warlike intentions) until the moment when the specialized forces which could only be aimed at one other player spring onto the map, ready for action, after a long period of secret building.

The best way to treat this type of system, which is incidentally usually in games in the very complex category, is on the same principle as reserves, discussed in the last chapter. Reinforcements appearing in the future *are*, conceptually speaking, reserves: you cannot use them now, and they may be required to face various situations then. It is usually unwise to gamble on the situation being as you expect when the fresh units become available, and the best thing is to produce forces for a variety of possible contingencies. The exception to this, which is frequently overlooked, is possible situations where you are going to lose anyway. There is no point in trying to cover a situation where all your neighbours attack you at once in a multi-player game, for instance, simply because you won't succeed – though it might be worth making it look difficult, in the hope of deterring them.

You should invariably try to maximize the number of possible situations where your long-term constructions will turn the game in your favour, even if this involves disaster in one or two conceivable outcomes. Here we may seem to part company from mathematical game theory. Suppose that I need 5 points to win a game, and my opponent, the wily Napoleon Clausewitz Smith, and I each have two possible courses of action, interacting to give me the points shown in the table:

		Smith chooses: Option A	Option B
I choose:	Option 1	0	4
	Option 2	5	2

Given this table, a game theorist would recommend a mixed strategy over a series of games, with each using the option some of the time. This maximizes the point score against best play. However, a player should be only interested in winning as compared with losing, and pay little attention to different degrees of failure (unless the game actually specifies different levels), so I would always go for Option 2, and hope that Smith had a brainstorm and took Option A, allured by the possibility of stopping me from scoring at all. In fact, there is no conflict with game theory, because, given that I don't wish to distinguish between different types of defeat, the values of the different outcomes to me, on a 0–10 scale, are:

		Smith chooses: Option A	Option B
I choose:	Option 1	0	0
	Option 2	10	0

Given this matrix, the game theorist would also opt for Option 2 throughout. There is simply nothing to lose. It may seem obvious, but it takes nerve in practice to gamble on the one remaining chance of success, if its failure would mean complete collapse, and another strategy is safe and gets close to winning; it is all too easy to think that the missing margin needed for victory with the conservative approach will turn up somehow, especially if the game is biased in its design to the other side. If there is a bias, it should be recognized from the start, so that one can be prepared to take greater risks to overcome it; conversely, playing with the bias, one should strive to keep everything as normal and risk-free as possible.

The basic requirement in all strategic questions is an ability to judge the position realistically, without false optimism leading to grand manœuvres which don't quite work out. Having decided the best plan with a reasonable chance of success, the tactical details rear their ugly head, and we shall look at these in the next part.

PART II

Tactics

Part I discussed the overall planning which players need to make during a game. In Part II, we shall look at the tactical operations which make possible the implementation of the 'grand design'.

Defensive terrain is considered first. No player should be indifferent to the type of ground on which his forces are fighting, and so the flow of battle will often be determined by the layout of the map, as was shown in some of the examples in Part I. In Chapter 6, we look at how to make terrain work best for the defender, and what the attacker can do about it.

Chapter 7 deals with the use of mixed forces, using two very different eras as illustration: the Second World War and the Thirty Years War. While the context and units involved differ radically, certain basic doctrines apply to both.

Chapter 8 focuses on naval and aerial rules, in particular in games where the main emphasis is on non-land combat, such as *Dreadnought* and *Midway*.

6 DEFENSIVE TERRAIN

Most wargames players like to be on the attack as much as possible. There is more glamour in a dramatic breakthrough with an armoured spearhead than in moving a battered rifle unit a hundred yards east to cover a weak spot in the sagging defence lines. There is more excitement when one rolls the dice to resolve attacks than when one methodically tries to ensure that the enemy has no good offensive options.

All this is perfectly natural and true as far as it goes. Yet good defensive play probably requires more skill, and withstanding a long series of assaults on every section of the line gives the defender a steadily growing feeling of quiet satisfaction. Let the other fellow rattle his rockets and tanks! 'I am armed so strong in honesty that your threats pass by me like the idle wind, which I respect not,' says Brutus in *Julius Caesar*; such is the attitude of the defender. There are quite a few players who always try to play the side mainly on the defensive, and one can recognize them by a certain calm and steadiness in their manner. One cannot envisage a Cossack cavalryman taking to defensive play.

Even the Cossack, however, would have to accept that every player needs a reasonable knowledge of defensive tactics, not least when he is trying to forestall them as the attacker. The nature of the mapboard assumes crucial importance in this context. The key to nearly every defence is the *intelligent use of terrain.*

We saw this from a strategic point of view in the *Stalingrad* example in Chapter 3. The Pripyat marshes forced the German to divide his forces into northern and southern groups, with all the accompanying dangers observed in that chapter. Later in the game, the river lines became a vital part of the Soviet defences. The tactical importance of the rivers, doubling defensive values, is enormous. It is scarcely an exaggeration to say that the success of a *Stalingrad* defence depends upon what percentage of his units during the game the Soviet player can shield behind rivers; only by doubling his strength in this way can he offset the German advantage in firepower.

Ultra-aggressive players are quite capable of ignoring such considerations and seizing chances for locally favourable counter-attacks in the open. The attacks often succeed, but the player usually loses; if the terrain is there for him to use, it will take unusually good odds to make it worth foregoing the advantage.

It is not quite accurate, however, to put forward the doubling of defensive strength as an end in itself. What is needed is sufficient defensive strength to make hazardous any attacks by whatever the enemy has got in the area. If the enemy is very weak, it may not greatly matter whether the defender is doubled or not – his opponent will not dare to attack anyway. If the position is being menaced by a large force, however, doubling may not be enough, and it may be necessary to pull back to a position out of reach of most of the enemy. The crucial factor may well be how many neighbouring hexes your opponent can get at, as it was in the Minsk dilemma in Chapter 3. Because it is difficult or impossible in most games to pass by an enemy unit without stopping, the attacker will not be able to get at a defender with flanking defensive forces on more than one adjacent hexagon. Given stacking limitations, this will restrict the amount of force that the enemy can bring to bear, and the position may be sound regardless of how many monstrous enemy behemoths are lurking menacingly in the area. In illustration 10 a detail is shown of a game of *NATO*. The West is attempting to hold on to Munich, hampered by the fact that some of the units shown here have only just been able to reach the area, while the 1-2-4s in Munich itself are local territorial forces restricted to the city. In *NATO*, all units have to stop on entering enemy ZOCs, with a few specialized exceptions not involved in this battle. The Soviet stacking limit is two units. Units attacked solely across rivers are doubled, as are defenders in town hexes (units with both advantages are still only doubled). The question in the picture is why the Western player has chosen to put his city defence units on the western edge of the city at 2447 instead of a hex further south-east, up against the enemy front line. Hasn't this simply handed half of Munich to the Warsaw Pact player?

The answer to this is that although eastern Munich *is* lost by this tactic, the city defence units are fairly safe from attack in western Munich. If they were moved up to the front, they could be attacked from three adjacent hexes across the river: 2549, 2647 and 2648. With six Warsaw Pact units in these three hexes, 28 attacking factors could be amassed, more than enough for a 3–1 on the doubled two units with two defence factors each. This would give the attackers immunity from 'attacker eliminated' results, a one-third chance of eliminating the defenders by forcing a retreat outside Munich, and a five-sixths chance of forcing eastern Munich anyway.

In the position shown, however, the city defence units can be attacked only from the single hex south-east of them (2545), on which only 10 Warsaw Pact factors can be accumulated, giving odds of 10–8 = 1–1, with a one-sixth chance of attacker elimination and still only a one-third chance of eliminating the defenders. If the Warsaw Pact player is sufficiently attack-mad to risk two armoured divisions trying to winkle out a couple of local defence brigades, by all means let him do so! Attacks over the river on the big flanking units are similarly perilous for the Soviet forces, and the only safe thing for him to do is to take out the exposed German 1-2-8 holding up the north end of the line. This will expose the 6-6-8 to attack on the following turn, but by then the Nato player will have gained time to reinforce the position, with the city defenders still intact. If reserves are short in the area, preserving these may make all the difference – at a pinch, the big units can cover open country and hope to survive, but if one had to be diverted to defend in Munich itself then there may not be enough material left to guard the flanks adequately.

The standard weapon against positional attacks is the *soak-off* attack, which in military terms is roughly equivalent to overrunning a lightly-defended position next to a strongpoint while keeping the main defenders occupied with a diversionary attack. The method is used in games where units are forced to attack enemy units to which they end their movement phase adjacent. The technique works when a weak unit and a strong one are defending an important position together, but can be attacked separately. The attacker makes sure of a crushing victory over the weak unit which will leave a large force next to the strong one; because the strong unit must also be attacked, a minor attacking force makes a sacrificial 'soak-off' attack on it. Very likely the minor attacker will be destroyed, but no matter. The strong enemy unit now has the main attacking force sitting on the edge of the river, fortress, or whatever the positional strongpoint may be. It therefore has either to withdraw (which accomplishes the aim of the attackers) or counter-attack. If it counter-attacks it cannot use the defensive advantage of the position, and will (if the original attack was well-planned) be forced to fight at unfavourable odds, resulting in a retreat and/or heavy losses or even elimination.

To give an example: if Moscow in *Stalingrad* is being defended by

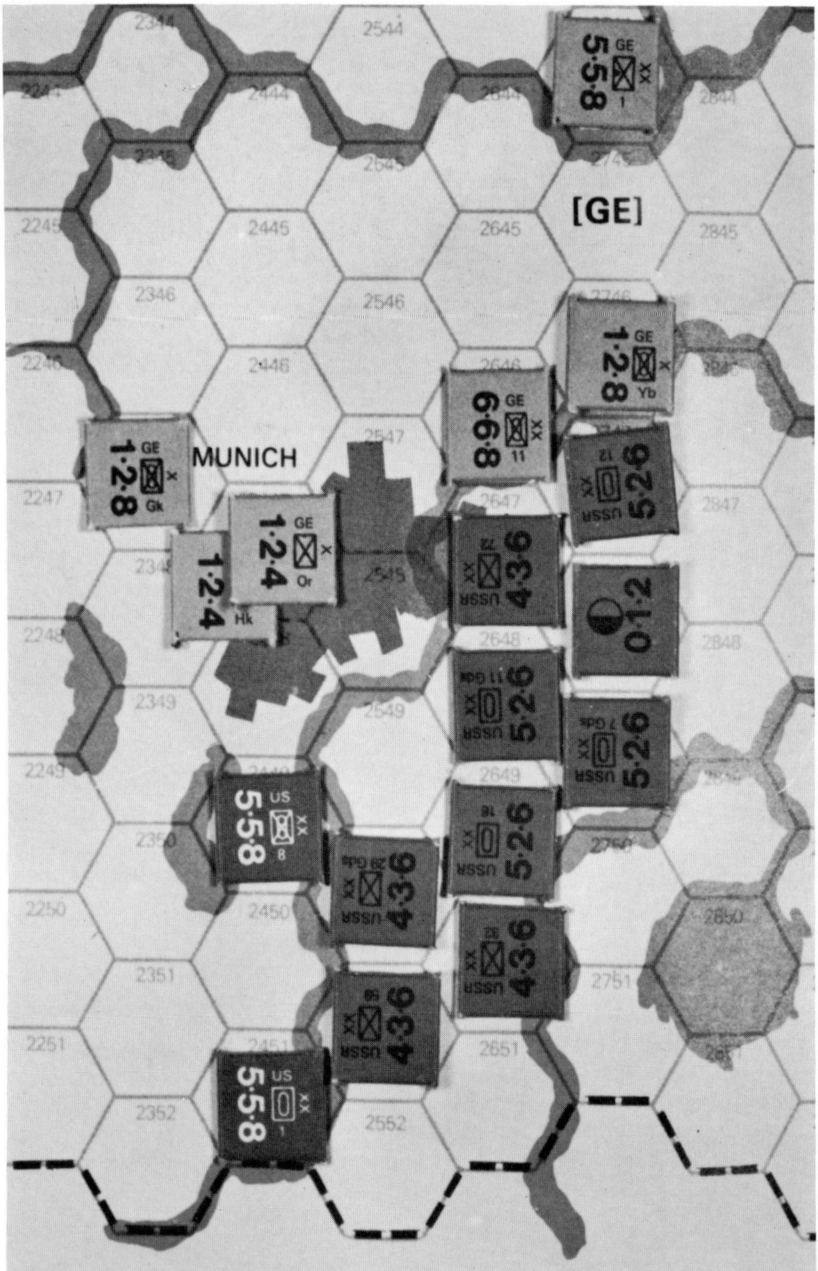

10 The Battle for Munich.

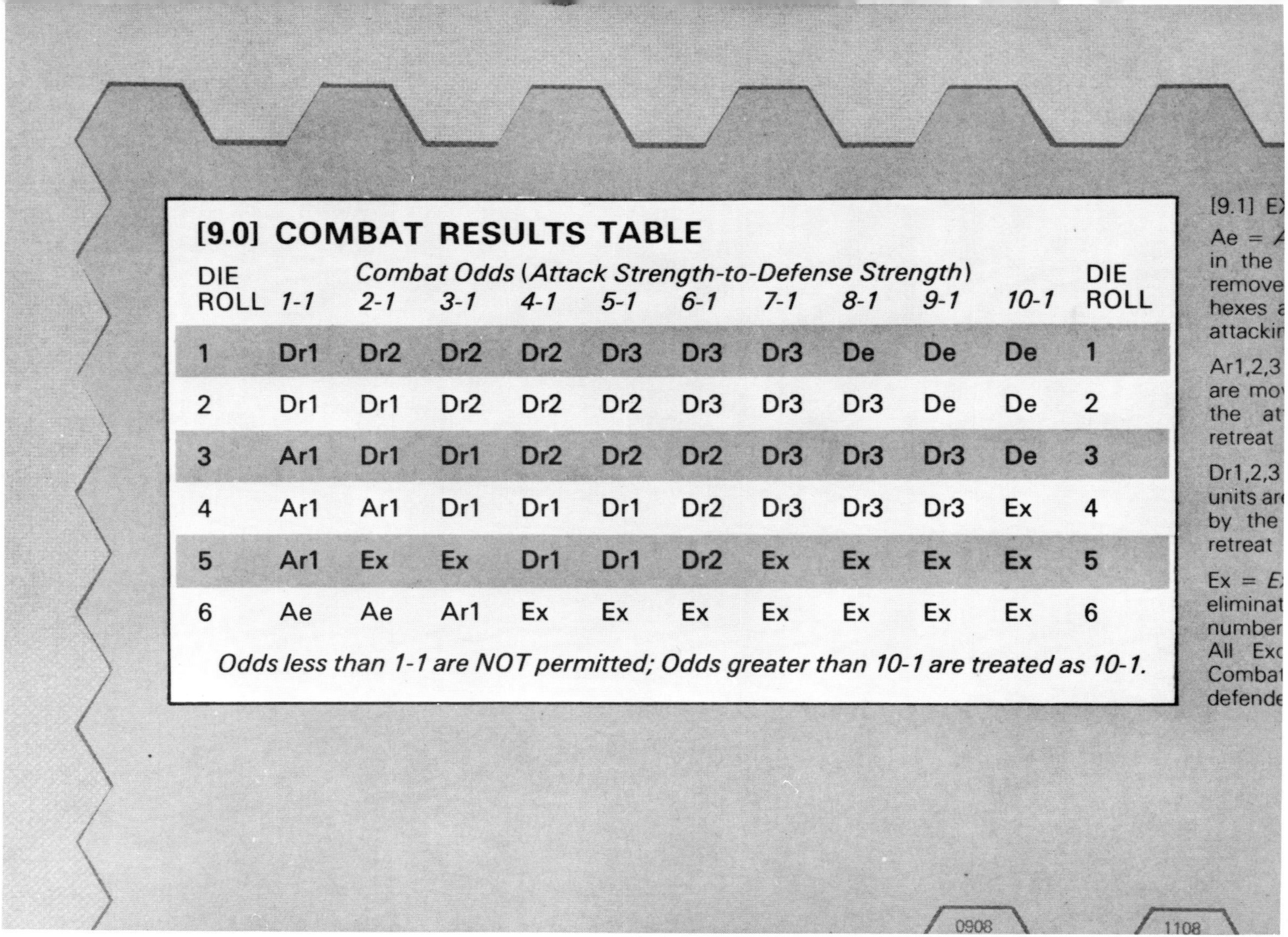

[9.0] COMBAT RESULTS TABLE

DIE ROLL	*1-1*	*2-1*	*3-1*	*4-1*	*5-1*	*6-1*	*7-1*	*8-1*	*9-1*	*10-1*	DIE ROLL
1	Dr1	Dr2	Dr2	Dr2	Dr3	Dr3	Dr3	De	De	De	1
2	Dr1	Dr1	Dr2	Dr2	Dr2	Dr3	Dr3	Dr3	De	De	2
3	Ar1	Dr1	Dr1	Dr2	Dr2	Dr2	Dr3	Dr3	Dr3	De	3
4	Ar1	Ar1	Dr1	Dr1	Dr1	Dr2	Dr3	Dr3	Dr3	Ex	4
5	Ar1	Ex	Ex	Dr1	Dr1	Dr2	Ex	Ex	Ex	Ex	5
6	Ae	Ae	Ar1	Ex	Ex	Ex	Ex	Ex	Ex	Ex	6

Combat Odds (Attack Strength-to-Defense Strength)

Odds less than 1-1 are NOT permitted; Odds greater than 10-1 are treated as 10-1.

11 CRT from *NATO*.

units with defence factors 10 and 3, then the total defence strength is 20 plus 6, requiring a massive 78 attacking factors to get a safe 3–1 on the two units together. If this is not conveniently possible, the German can simply attack the two separately:

MAIN ASSAULT, *v* defence factor 3: 18 attack factors, at odds 18–(3 × 2) = 18–6 = 3–1.

SOAK-OFF DIVERSION, *v* defence factor 10: 4 attack factors, at odds 4–(10 × 2) = 4–20 = 1–5.

The first attack will dislodge the 3, leaving either 12 or 18 attackers from that battle (depending on whether or not an 'exchange' was rolled) next to Moscow. The second result will result in the attacking 4 being retreated or destroyed. Who cares? On the Soviet turn, the 10-factor unit, which incidentally will only have 7 *attack* factors, is faced with a choice of pulling out of Moscow or counter-attacking at 7–12 or 7–18, i.e. 1–2 or 1–3, both of which will probably force him to retreat anyway, if he survives. In practice, therefore, he will probably retreat of his own accord (assuming no reinforcements can be rushed in), and Moscow will have fallen to a mere 22 attack factors instead of the 78 which seemed to be necessary. Thus the 3-factor defence unit was just a nuisance for the Soviet player, getting in the way of an orderly defence: on its own, the 10-factor force could not have been safely assaulted with under 60 factors. The realism of this is disputable, and games in which attacks are not compulsory are increasing in number, but there are plenty of both kinds, and it is vital to know the soak-off technique in games of the more traditional type.

In practice, the soak-off technique is rarely used to such effect as in the above example, because the defender wasn't born yesterday either. The defence against the technique is simple: don't mix weak and powerful defenders in the same sector. If both defenders have the same strength, then the soak-off loses much of its sting, as the *victorious* attack against one defender requires more attacking strength than may be available. No longer can the defensive strength of one unit be neutralized by the fact that its accomplice is a pushover. Units of different strength should, as far as possible, be kept sufficiently far apart that they cannot be attacked from the same hex. Even if they are not stacked together, a soak-off can be equally effective if the forces successfully attacking one are in a hex next to the other.

Whether a game uses the controversial attack rule or not (and there are arguments in its favour: is it reasonable that you can sit out your phase without any conflict with a hostile force in the next hex?), the defender will have a special incentive to fight hard for good positions.

Generally speaking, he will be the weaker party at that stage in the game, and will have to reckon that the enemy will be hunting him for a decisive battle as soon as possible, before friendly reinforcements arrive or time runs out. If there is going to be a battle with unequal odds, the only way to even things up is to cling tenaciously to positions giving defensive advantages. It follows that it makes more sense to gamble and take risks when victory would mean the recapture or continued possession of a key strongpoint. In the above example, the Soviet player might decide to counter-attack at 1–2 if he were sufficiently desperate. In the open, his 10-factor unit would stand little chance. If he could hold on to Moscow, then perhaps the tide could be stemmed until reinforcements arrived.

This line of reasoning is all very well, but it is used far too often, to justify holding on to some unimportant doubled position with a dangerous counter-attacking manœuvre, when it is about to be outflanked anyway. We are back with the 'attacking psychology' of players here. Give the average player a halfway decent excuse to counter-attack, and he will seize upon it with indecent alacrity. Take the *NATO* example earlier. If a counter-attack were necessary to set up the line shown, for instance by attacking a forward Soviet unit which might have slipped over the river, then this would make good sense. Not so if there were a Soviet unit in eastern Munich, and an attack on it was risky, because it involved poor odds or weakening another part of the line by diverting reinforcements. As we saw, there is no reason to fight for eastern Munich, despite the fact that it is a doubled position, because defending it is unwise even if it is actually empty at the moment. Forcing out a Soviet unit may be very satisfying but in this position it doesn't make rational sense.

There are two criteria to examine in deciding whether a position is so important as to be worth taking risks for:

1) How long can it be held if the counter-attack succeeds?
2) Is it the key to other strongpoints?

The second criterion is especially relevant to river lines, as the defensive bonus will fail in other river positions if it can be crossed at one point and supporting attacks pushed along the defending bank. If there are other worthwhile river positions, then the defender may be inclined to put up a fight to protect them, either by counter-attacking the lost position in an attempt to restore the full river line, or by defending the sector in the open (for once) if this can be done in a manner protecting the remaining doubled river defenders from being outflanked and undoubled.

All these considerations lead inevitably back to the victory conditions. A player on the defensive should have time on his side, unless the game is unusually unbalanced, to compensate for his tactical inferiority. Most games have a time limit, and it is virtually always the stronger side who have to achieve certain goals by then, as if it were the weaker side, then the stronger force could simply sit tight and let the turns roll by until victory. In many games, of course, there is no clear-cut distinction between stronger and weaker sides, with both players having periods of dominance according to terrain and the flow of reinforcements; nonetheless, at any given moment, there will normally be one player on the offensive and the other wishing the game-clock would get a move on. Playing for time involves careful planning ahead, as discussed in the earlier chapters, and it also affects the decision as to whether a particular strongpoint is worth holding to the last gasp. In the *Stalingrad* example, it is worth taking a big risk for Moscow, because it is a replacement city, yielding a steadily increasing flow of fresh troops as long as it is kept from the Germans, and because the tide in the game turns to the Soviet side if the invaders have been kept at bay for the first twelve to eighteen months. In the *NATO* case, on the other hand, there is no decisive swing in the later stages. The reinforcements on both sides have virtually all arrived when the fifteenth turn is played, and any Soviet weakness towards the game end is mainly caused by the Warsaw Pact player having bled his forces white trying to seize the cities, which are his territorial objectives. In the position illustrated, it does not look as though Munich can be held for very long, so it is not to be defended 'at all costs', and should just be used as a means to force a heavy price in Warsaw Pact losses. Similarly, there is no reason for the Warsaw Pact to rush unduly; they should try and outflank the position before seizing it with the cheapest frontal attack possible.

A striking example of the lunacy of fighting for everything as if it represented the last spot on earth is provided by optional rules in SPI's *Turning Point* (on the tactical battle for Stalingrad), showing what happens when the German units are forced to follow Hitler's order, 'Where a German soldier stands he will not retreat.' This swings play balance sharply to the Soviet player, compared with the standard rules where the Germans are allowed to withdraw when it seems the sensible thing to do.

The reason why defence generally requires more skill than attack is that it will more frequently be disastrous if one or two units are misplaced. The ultimate, cardinal and unforgivable sin is to leave a hole in the line, and unlike most sins retribution is usually instantaneous and terminal. Anyone making a habit of this should seriously consider switching to Snakes and Ladders (fortunately there is no need to worry about this – after a few games, a gap in the lines sticks out at a glance). Without doing anything terrible like this, it is only too easy to overlook an enemy concentration of forces requiring a strengthening of the defences in that area, or to fail to allow for all the possible results of a counter-attack, from a triumphant advance (leaving a surroundable unit with a gap behind?) to an unexpected repulse (dislodging the flanking protection for a vital river position?). If there is a second movement phase after combat, this makes life easier, but it is still important to check that reinforcements are available to stop any gaps which combat may have opened. Most nerve-wracking of all are the games (mostly tactical ones, e.g. *Sniper* and *Air Force*) with simultaneous movement, though in some ways this approach gives a certain possibility of getting away with errors, because your opponent never dreamed you'd do something silly like *that*, and is not in a position to exploit it. If you think your move was eminently sound, and do nothing to correct it, then that, of course, is a different matter!

A more or less watertight defence takes a little while to check, though like most aspects of the games this speeds up with practice. The ideal defence should be built up along the following lines, with checking of earlier steps periodically to make certain that the conclusions reached earlier have not been invalidated by later discoveries.

1) What is the general area where a stand should be made this turn?
2) Has the other player left any tempting targets – exposed units, recapturable objectives, or even (special Christmas present) a hole in the line which you can penetrate to encircle his units?
3) For *each* hex you want to hold: what force can the enemy bring to bear on it, and can you defend with sufficient strength to make an attack by him dangerous or expensive for his forces?

If the calculation leads to an unpalatable conclusion, then steps 1 and 2 need to be reconsidered: the area is indefensible, or the counter-attacks too ambitious.

If all possible plans look equally unsatisfactory from the point of view of holding a solid line, then restrict the evidently inevitable enemy advances to as small areas as possible, because this will mean that most of the defence can be used again next time without ceding a lot of ground. A large force crossing a river at a single point is much easier to contain than little penetrations all along the line. If you have to choose, it is probably better to make the line solid everywhere except

12 The Monschau Road. Standard US setup.

TOURNAMENT GAME BATTLE RESULTS TABLE

1-6	1-5	1-4	1-3	1-2	1-1		2-1	3-1	4-1	5-1	6-1	7-1	8-1
A elim D advance 1	A elim D advance 1	A elim D advance 1	Contact	D back 1 A advance 1	D back 2 A advance 1	**1**	D back 2 A advance 1	Exchange A advance 1	Exchange A advance 1	Exchange A advance 1	Exchange A advance 1	Exchange A advance 1	D elim A advance 1
A elim D advance 1	A elim D advance 1	Engaged	Engaged	Contact	Contact	**2**	D back 1 A advance 1	D back 3 A advance 2	D back 4 A advance 3	D back 4 A advance 4	D elim A advance 1	D elim A advance 1	D elim A advance 1
A elim D advance 1	A back 1 D advance 1	A back 1 D advance 1	A back 1 D advance 1	Engaged	Engaged	**3**	Contact	D back 2 A advance 1	D back 3 A advance 2	D back 4 A advance 3	D back 4 A advance 4	D elim A advance 1	D elim A advance 1
A back 2 D advance 1	A back 2 D advance 1	A back 2 D advance 1	A back 2 D advance 1	A back 1 D advance 1	Engaged	**4**	Engaged	D back 1 A advance 1	D back 2 A advance 1	D back 3 A advance 2	D back 4 A advance 3	D back 4 A advance 4	D elim A advance 1
A back 3 D advance 2	A back 3 D advance 2	A back 2 D advance 2	A back 2 D advance 2	A back 2 D advance 1	A back 1 D advance 1	**5**	Engaged	Contact	D back 1 A advance 1	D back 2 A advance 1	D back 3 A advance 2	D back 4 A advance 3	D back 4 A advance 3
A back 3 D advance 3	A back 3 D advance 3	A back 3 D advance 2	A back 3 D advance 2	A back 3 D advance 2	A back 2 D advance 1	**6**	A back 1 D advance 1	Engaged	Engaged	Contact	D back 2 A advance 1	D back 3 A advance 2	D back 3 A advance 2

Odds worse than 1-6 are not allowed. Odds greater than 8-1 are treated as 8-1.

EXPLANATION OF TOURNAMENT GAME BATTLE RESULTS

ELIM: Same as in Basic Game.

BACK: Same as in Basic Game with the following important addition: whenever possible, losing Units must be retreated to and/or along the nearest road. Where there is more than one road equidistant, and at intersections, the loser has the choice of retreat route.

ADVANCE: After losing Units have been retreated, the winner has the option to advance all victorious Units up to the number of squares specified. (Example: at 5-1, a die roll of 3 means that all defending Units are retreated 4 squares and all attacking Units may advance 0, 1, 2 or 3 squares.) Units may advance in any direction according to these restrictions: (a) the first square of advance must be the loser's vacated square, and (b) advancing Units must stop as soon as they land on an enemy controlled square, and (c) Units may advance directly into enemy controlled squares **only** if no alternate advance routes are available. Units that advance adjacent to enemy Units whose battles have not been resolved do not participate in those battles . . . however, such placement does serve to cut off retreat routes. The winner may advance the full number of squares even when losing Units are eliminated because of blocked retreat routes.

EXCHANGE: Same as in Basic Game.

ENGAGED: Essentially the same as in Basic Game; however, both players are allowed to reinforce. Thus, the routine is as follows:

STEP 1: The defending Units, only, are turned upside-down.

STEP 2: The defender, in his Turn, cannot move his upside-down Units. He does not have to counter-attack although he has the option to do so. Also, he does not have to attack non-engaged enemy Units his Units have remained adjacent to.

STEP 3: The attacker, in his following Turn, cannot move his engaged Units and must attack, again, the upside-down Units he attacked in his previous Turn.

Defender's options in STEP 2 above:

(a) He may counter-attack with his upside-down Units, reinforcing them with new Units who also attack. The upside-down Units are turned right-side-up and normal combat procedures are followed. In this event, all of the opponent's engaged Units are freed from their obligation to attack again as in STEP 3 above.

(b) He may reinforce his upside-down Units by bringing up additional Units and turning them upside-down also. Reinforcements can only be placed on squares already containing upside-down Units, subject to the 3-high stacking limitation. He does not counter-attack in this option.

Attacker's options in STEP 3 above:

(a) If his opponent has exercised option (a), his engaged Units are freed from their obligation to attack again. In this event, he may move his freed Units in the normal manner. He may, however, attack again if he wishes.

(b) If his opponent has exercised option (b), he must attack all upside-down Units. He may bring up reinforcements and attack with them also.

IMPORTANT: In all counter-attack situations, players must abide by the **Multiple Unit Battle** rules which require the attacker to fight all adjacent Units. In cases where engaged Units are adjacent to non-engaged enemy Units, the attacker has the choice of dividing combat in any manner as long as all adjacent enemy Units are attacked in one way or another. (See diagrams in the Battle Manual.)

CONTACT: No casualties are taken and there is no retreat or advance. But the defender, in his Turn, must either withdraw or counter-attack.

602

13 CRT from *Battle of the Bulge.*

at one or two points, where the defence is nothing but a cheap delaying unit or two, rather than allow your opponent the luxury of a number of rewarding attacks at different points.

We considered the problem of reserves in the strategy part of this book. This has tactical implications as well. Illustration 12 shows the starting position of the northern US defence line in *Battle of the Bulge.* The general situation at the start of the game is that the Germans are pouring through defensive gaps to the south, hindered only by terrain and small groups of US defenders. Here in the north, their progress will be slow, with most of the terrain rough, and advance only really possible along a single, easily defended road leading south-west from the front line. If a rapid advance were the only thing that interested the German player, he would be well-advised to use most of his troops in more rewarding terrain further south. However, there is a good reason to commit a sizeable force to the north: the two American divisions (six 4-4 units). If these are able to move south, one can be sure that they will be rushing to the rescue of their colleagues on the very next turn, leaving just one or two units to hold the road. But the road can be blocked to the American regiments with just a small German advance, and then the horrible terrain suddenly works in favour of the Germans, with a large US force cut off in the important turns to come. Moreover, the *Battle of the Bulge* CRT has an unusual feature: it is possible to get 'engaged' as a result, which means that the defender can only withdraw from the battle if it is an armoured unit; otherwise, it must counter-attack or stay put (in this case it is not *required* to attack, though attack in enemy ZOCs is otherwise compulsory in the game) waiting for the next enemy assault. As a counter-attack will surrender the defensive advantage of the rough terrain, against an enemy strong enough to have attacked the defender doubled, staying put is the most likely US choice; the US regiments shown are all infantry, and no reinforcements are available. Thus every engaged result will pin the defending unit. If the positions around it are crumbling then it can probably be surrounded and attacked with massive force on the following turn, wiping it out altogether. This is very pleasant for the attackers, but it is important not to lose sight of the main target: stopping the regiments from slipping south.

Problem: Given the German forces shown, all of which can reach anywhere in column SS on the map, in stacks of up to *three*, find the optimum attack. The CRT is shown in illustration 13. A 'contact' result compels the defender to withdraw or counter-attack in his turn; for the reasons mentioned in the discussion of engaged results, he will normally choose to withdraw in this position. Units within one hex of a road unblocked by enemy ZOCs can move to the road and zip along it, on the same turn; apart from this, units moving into a rough terrain hex must stop. There is a road running south from the line of US units in the column RR; the US player will be even happier to get his forces away down this road, which is direct to the area where he needs reinforcements.

Attacks only from units on rivers are halved, as are attacks on rough terrain or towns. Every US unit with a German unit next to it must be attacked by someone. Retreats are made towards roads. Advances of one hex *through* enemy ZOCs are allowed if dictated by the CRT, but after that the advance is stopped by a ZOC.

When comparing your solution to my suggested plan in Appendix A, the first criterion should be how many of the six units are likely to be tied up (including any needed to block the south-west road from a German breakthrough), and the second criterion what prospects there are of speedy eliminations. It is taken for granted that you won't leave a gaping hole for the US to surround *you*! Unless you do something like this, you can safely assume that the US player will be in no mood to counter-attack. On your next turn, you can expect to have fresh units which have worked round and can come up the road from the south leading to the current US line in column RR.

Note that from north to south, the US units on defence are respectively: doubled (Monschau town), doubled (rough terrain), undoubled, doubled (rough terrain) and undoubled. The southernmost of these five hexes has *two* 4-4 regiments on it, and these must be treated like an 8-4, i.e. attacked together or not at all.

7 COMBINED ARMS

One of the most interesting aspects of tactical games is the wide variety of unit types. In a purely strategic game, there will not normally be room for much differentiation of forces, without making matters unbearably complicated. At the tactical level, therefore, a new skill is called for: the ability to build powerful and effective strike forces out of a bewildering array of units of different kinds, and to use them in combat in the optimum way.

One of the pioneering games in this field was *Panzerblitz*, set in an anonymous part of the Soviet countryside during the Second World War. The basic rules for dealing with the fifty-seven German and Soviet unit types are used with a chameleon-like three-part mapboard which can be joined up in numerous different ways to simulate a dozen different local battles, ranging from small local skirmishes to the enormous battle of Prochorovka with a Soviet Tank Corps facing a German SS Panzer Division. With its west-front companion *Panzerleader*, *Panzerblitz* is still one of the most frequently played games around, despite some criticisms and various new games at the same level since. We will use the game to provide illustrations for some of the concepts in this chapter.

The units on each side in the game can mostly be fitted into four general categories: armour (tanks, tank-destroyers, armoured cars, self-propelled artillery and assault guns), infantry, towed guns, and transport (half-tracks, trucks and slow-moving wagons). Each of them is given four factors, which appear on the counters (see illustration 14): top left is attack factor, bottom left defence factor, top right range, and bottom right speed. Note in the illustration how slow the infantry and towed guns are on their own; some guns cannot move at all without transport, which takes a stationary turn to load them up. The letter (top centre) on the counter shows the type of weapon:

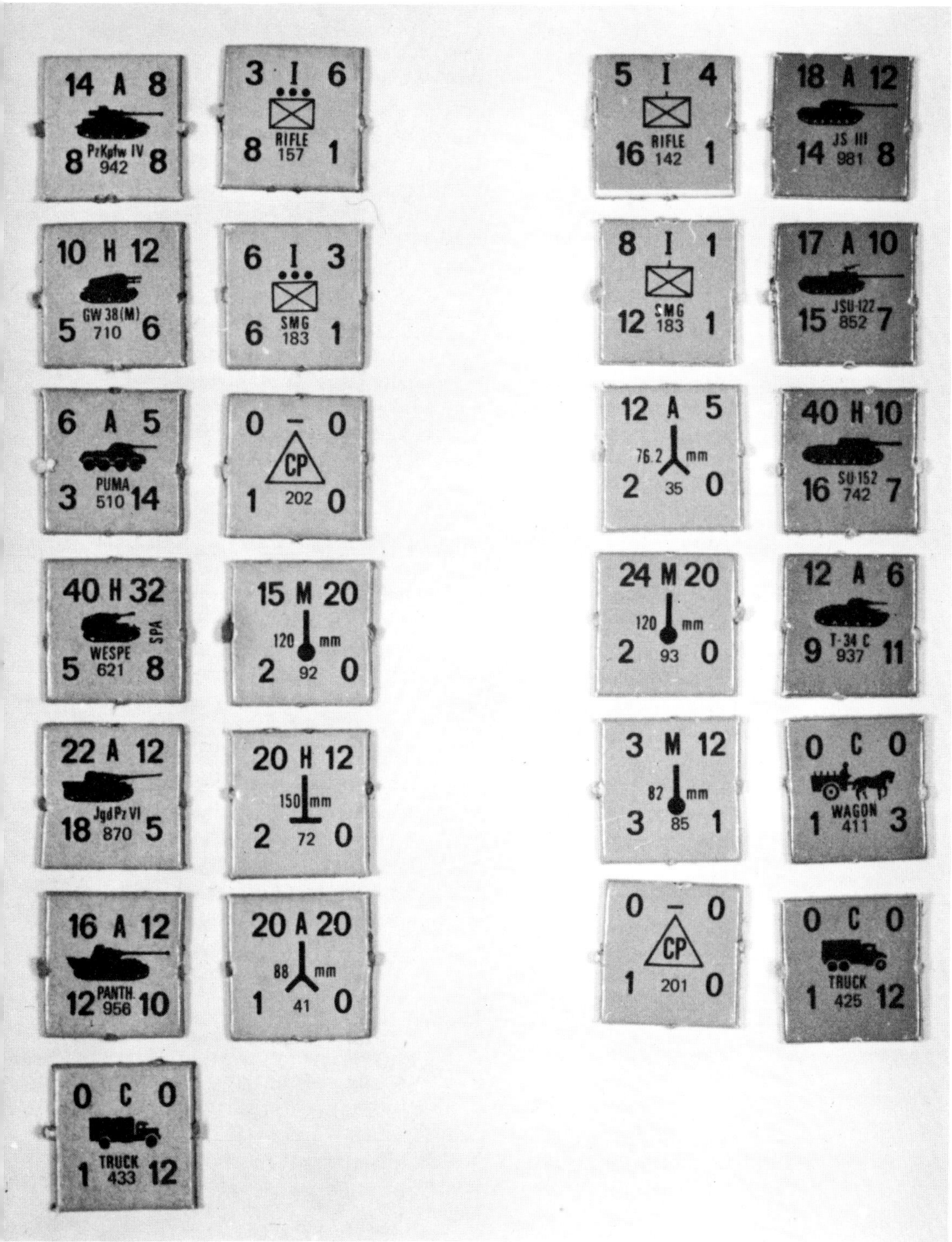

14 The *Panzerblitz* troops.

A class weapons are armour-piercing, and have double strength at armour under half range, but only half strength against non-armour.
H and M class weapons (howitzers and mortars) are halved against armour over half range.
Infantry cannot attack armour at all, except by physically moving up to it on the next hex, in which case they get an improvement on the die-roll to compensate for the otherwise generally feeble infantry attack factor compared with most tank defence strengths.

Terrain also affects movement and fire, the most important point being that units in towns and woods, of which there are quite a few, cannot be fired upon unless there is a unit in the next hex to 'spot' (guide in) the shots. The numerous hills of different heights block fire when they get in the way, unless a special observation unit (CP, standing for Command Post) is in view of the target, and the firing unit is a mortar or self-propelled artillery.

The stacking limit is three for Germans, two for the Soviet forces.

This variety of capacities naturally gives the players a good deal to think about on each turn, as they can never be sure what will be thrown at any particular position, or what opportunities will be available for exploitation in the enemy positions. If one unloads a large force of infantry in what appears to be a key town, he may find that the other sends his forces speeding past to some objective behind, wreaking appalling havoc before the infantry can get back into its transport and come to the rescue. But infantry and guns still loaded in transport are not permitted to attack, so if you delay too long in committing your troops you may find that they have missed a series of opportunities.

The general guidelines for combined arms in modern wargames are these:

Armour is designed for rapid penetration in strength. Except for the few slower units in this category, which use their movement factors primarily to move to good firing positions for several turns, armour should always be prepared to move in any direction, to exploit whatever point in the enemy defences your opponent may have weakened on his previous turn. In a game like *Panzerblitz*, which is dominated by terrain, you should systematically decide which defensive positions you want to capture, each with a view to establishing a powerful firing position against the next. At all costs avoid committing armour in tiny, fragmented groups; *every* type of unit is most effective against armour at close range, and nothing will please your opponent more than if you assault his forces each turn with just enough units for him to knock out. Each enemy unit can only fire once per turn; overload the defences so that there are some of your forces which he cannot attack effectively, leaving them to use *their* close-range effect (and, in the case of *Panzerblitz*, ability to spot for long-range artillery) to destroy the defenders next turn. Frequently in tactical games a slow-moving unit will be unable to spot anything to shoot at; to offer it a target which it can knock out is absolute lunacy – yet it is the most common error in this type of game. It always pays to look ahead a bit, but in this sort of situation it is essential: if you want to assault a defensive position, work out what the enemy is going to be able to do about it. If he can probably knock out the attackers, then wait until reinforcements build up to give a better chance of overloading his units.

This worthy advice may be easier said than done if the game has simultaneous movement, like *Kampfpanzer*, one of SPI's alternatives to AH's *Panzerblitz* group. If both players are moving at the same time, then each will have to guess the enemy plans in the light of his former moves, opening the possibility of feints disguising the real intentions. This has obvious advantages in realism and excitement, but it has drawbacks as well: it can be simply impossible to guess what the opponent will do next, and if the movement period simulated is substantial (i.e. if the position is going to change markedly during the move), the inability to predict anything makes the game rather too chance-dominated. From the point of view of realism, the best solution may be to set the length of movement period simulated equal to the time it would take to issue fresh instructions. Instead of having alternate movement and fire phases, as in most games, there would be a series of short movement phases (slow units would only move in some of them), allowing the players to react to enemy movements; as a rifle platoon edged through a wood and saw a tank group bearing down on them, they would drop flat instead of wandering blindly down into the path of the jubilant tank-commanders. Like all tactical games, this would work best with hidden movement, allowing units to sneak up on enemy positions. The *Kampfpanzer* scale, incidentally, is 220 seconds per move.

After this digression on game design, to return to the discussion of the roles of the different unit types:

Infantry: the primary role of infantry is defensive and close combat. Once committed, it tends to stay in the same area, as it is difficult, hazardous and slow to pick up troops from the front line and whisk them off somewhere else. In *Panzerblitz*, infantry alone can move and fire on the same turn, in the form of the close-assault technique described earlier (armour can also execute combat *during* movement, by overrunning enemy positions, but this is only when the enemy has left something in the open, which is rare unless your opponent is of a suicidal disposition). In view of this special role of infantry, they are particularly suited to attacking woods and large towns, in which they can be dropped on one turn (many *Panzerblitz* players are unaware of the important ruling that transport can move away after dropping off units on the same turn) in a 'safe' end of the area, and then move forward with guns blazing on subsequent turns. Isolated patches of wood, and small towns are much more difficult to attack, because the attackers have to be dropped outside the covering terrain to get at the defenders on the following turn. For these cases, it is necessary to return to the 'overloading' technique used by armour.

Towed guns: these are generally pretty useless in *Panzerblitz* (except hidden-movement versions) because of the so-called 'Panzerbush' syndrome, in which all units dodge nervously from bush to bush, and can never be spotted for fire except when they are forced to assault a position from the open. If you have a nest of guns somewhere, you can be sure that this will not be the bush where your opponent chooses to go, and as the guns are even less mobile than infantry, as a rule, there is little to be done with them except to assign them to static defence of places which you really want to hold throughout the game, preferably on hilltops from where they can give assistance to any of your other positions in the area if they need it. In other games, and in particular those with hidden movement, towed guns can be quite deadly, but their strength is nearly always in defence, preferably in the form of an unexpected ambush against onrushing enemy forces. Guns which can be guided by the CP spotting units are a bit better than the rest, because they can all concentrate on whatever the CPs can see on any given turn. This type of long-range artillery should be tucked away somewhere safe, with heavily-guarded CPs manning strategic hilltops. The tremendous value of CPs is rarely fully appreciated, probably because they look so puny.

Transport: in many games, including this one, some or all infantry and gun units can go into battle carried or towed by armoured units; if so, this reduces the number of pure transport units required. In general, transport should stay close to 'foot' units in case they need taking elsewhere. Transport should be *fast*; if you must have wagons or some other slow vehicle, put your least important passengers in it, so that the main force can be rushed in wherever it may be needed. In *Panzerblitz*, trucks are often used to help overload a defence, as they (rather unrealistically) can guide incoming fire; this technique is extremely useful, as trucks are often pretty superfluous once they have dropped the infantry where it wants to go. It is also a sure-fire way of infuriating your opponent, as few things are more annoying than having three Panther platoons surrounded by four truck companies, with the survivor of their withering fire bringing in a torrential artillery barrage and destroying them.

It is usually sensible to have a 'little bit of everything' in most of the groups, to cope with the different types of enemy which they may be required to deal with. Infantry riding in on tanks can be particularly effective, as they dismount on reaching the target, adding their defensive power to the forces overloading the position, and giving the defender a strong incentive to evacuate the position – as infantry is frequently unable to find anything to attack in its vicinity, your opponent will be in the habit of trying to keep it unemployed by staying out of its way. Unsupported armour is vulnerable at close range to armour-piercing guns at doubled strength, and in view of its splendid mobility you will not want it to hang about occupying captured positions.

Problem: Below are the values of the units shown in illustration 7a, using the 'Situation 13' point system, devised by Tom Olesen to allow players to select their own forces (*see* Appendix C). You can, for the purposes of this problem, choose as many as you like up to eight of each unit, provided you count the value of each. Choose 600 points' worth from each side, and divide them into groups. Assume the German objectives are:

(*a*) capturing a large wood
(*b*) breaking through enemy lines to his undefended rear areas
(*c*) holding onto a small (single hex) village at a crucial road intersection in open terrain.

Assume the Soviet objectives are:

(*a*) holding the wood
(*b*) preventing the breakthrough
(*c*) capturing the village.

Which group will you assign to each task? If you know someone else interested in wargames, get him to choose forces as well, and compare the opposing groups, his Russians against your Germans and *vice versa*. There are various reasonable solutions; the one which I would choose, together with a guide to predicting the outcome of each clash according to what you chose, can be found in Appendix A.

The distance between the front lines is 9 to 12 hexes, flat and open country, with scattered cover for each side's units at the front. High ground for artillery behind the lines is available. You can expect to disable enemy units for a turn if you attack them at 2–1 or better, and destroy them at 4–1 or better; less than 2–1 makes it rather doubtful (though 1–1 in close assault should work).

To avoid a solution dependent on the peculiarities of a particular position, and keep to general principles, this problem is deliberately abstract, and could not in fact arise on the *Panzerblitz* boards. As all the actions will be going on simultaneously, you should assume that units committed to one action will not in general be available for another, though it is worth bearing in mind the possibility of switching some of the units to other groups if the need arises.

In the table below, AF = Attack Factor, DF = Defence Factor, R = Range, MF = Movement Factor.

SOVIET UNION

Unit	AF	DF	R	MF	Cost
76·2 mm (A) guns	12	2	5	—	10·5
82 mm (M) guns	3	3	12	1	11·5
120 mm (M) guns	24	2	20	—	24
Rifle infantry	5	16	4	1	23
Submachinegun (SMG)	8	12	1	1	22
CP	—	1	—	—	5
Wagon	—	1	—	3	4
Truck	—	1	—	12	7
SU152 (H)	40	16	10	7	68
JSU122 (A)	17	15	10	7	49
T34c (A)	12	9	6	11	38
JS111 (A)	18	14	12	8	52

GERMANY

Unit	AF	DF	R	MF	Cost
150 mm (H) guns	20	2	12	—	18
88 mm (A) guns	20	1	20	—	21
120 mm (M) guns	15	2	20	—	19·5
Rifles	3	8	6	1	13
SMG	6	6	3	1	14
CP	—	1	—	—	5
Wagon	—	1	—	3	4
Truck	—	1	—	12	7
Puma (A)	6	3	5	14	28
Wespe (H)	40	25	32	8	69
Gw38 (A)	10	5	12	6	27
JgdPzVI (A)	22	18	12	5	57
PzKpfwIV (A)	14	8	8	8	38
Panther	16	12	12	10	50

All this would be a bit specialized for inclusion in this book, were it not for the fact that the basic techniques of combined arms tend to reappear in any games of the type, whether the subject is the armoured struggles which we have just studied or something quite different, like the duels of cavalry, archers and elephant troops in *Alexander*. A few principles are true of all games with a variety of unit types, not just tactical ones:

1) When placing your forces initially you should be prepared for unexpected developments. Try to have specialized units in each area where their expertise can be put to use, rather than mostly grouped together in one area where they may not be really needed in the event. In several games, for instance, rail or road lines can be laid by special

engineer units (*Jerusalem* and *Burma* are good examples). At the start, one may think one can see where the units will be needed, but only in some games can you be sure, and if not then one is well-advised to keep a reserve to allow for the possibility of one's opponent thinking of something unexpected.
2) On the other hand, establish by practice what is the minimum force of any type to be useful. Asking a single infantry unit in *Panzerblitz* to do anything is probably over-optimistic; in other games (e.g. *Third Reich, NATO*) it may be important to group your armour together, fly bombers in groups for mutual protection (*Luftwaffe*), or sail in a tightly formed squadron of ships (most pre-steam naval games).
3) Use high movement factors for positional manœuvre rather than merely accelerating the battle. If some of your troops move faster than the enemy, then they can often keep your opponent guessing, and force him to guard several points on which they might swoop. Once they are committed, this advantage is lost, even if the skirmish itself seems to be paying off.
4) Mix high and low *combat* factors in groups with similar movement allowances. In different situations, different units will go to the front: strong forces for major attacks, weak ones for screens, and probes intended to lure out an enemy reaction.
5) In short games, you can actually count hexes to see where your different units will be able to reach by game's end. When in addition the objectives are territorial as in *Winter War*, the calculation may make the difference between an initial setup with a chance of success and the reverse, as anyone who has tried to march the Soviet army in that game south from Petsamo will agree.

In some games, a particular class of unit may be so powerful as to decide the outcome, if it is possible to use it to full effect. There is always a fly in the ointment: it is so slow that it can't reach its targets easily (Gatling guns in *Custer's Last Stand*), or it only works against targets in the open (*Panzerblitz* overruns, *Seelöwe* air raids) or already weakened (cavalry charges in *Thirty Years War* Quad games). Initial placement of these should be made with especial care; decide first where these are going, and then build your strategy and the placement of other units around this decision. Any balanced game of this type will be based on the assumption that the special capabilities will be available free of drawbacks only part of the time. If you can circumvent the problems, you will take a substantial stride towards victory before a shot is fired.

As an example of special unit functions, illustration 15 shows the

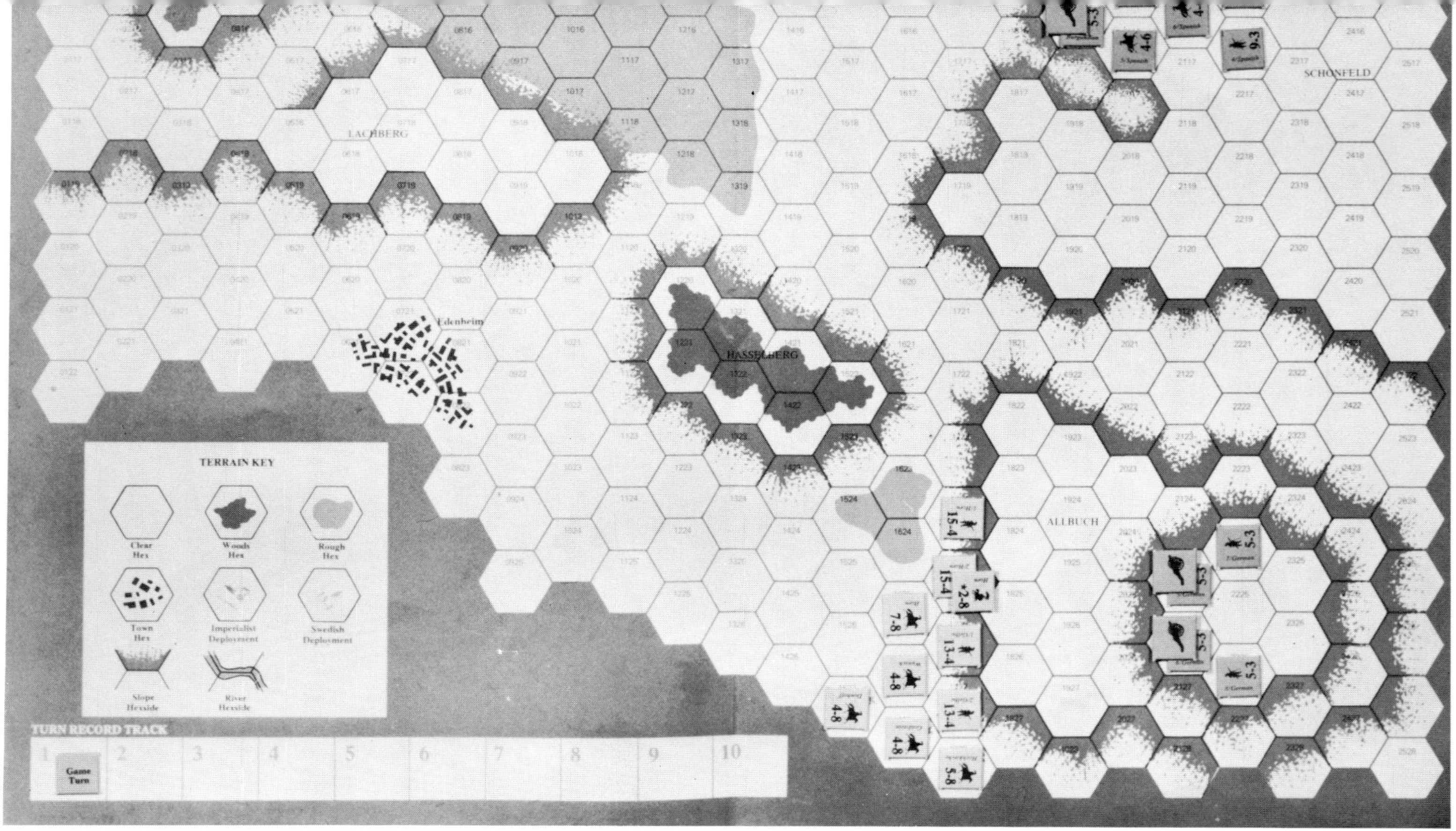
LACHBERG
Edenheim
HASSELBERG
ALLBUCH
SCHONFELD
TERRAIN KEY
Clear Hex
Woods Hex
Rough Hex
Town Hex
Imperialist Deployment
Swedish Deployment
Slope Hexside
River Hexside
TURN RECORD TRACK
1
2
3
4
5
6
7
8
9
10
Game Turn

15 The Battle of Nordlingen.

setup from a game of *Nordlingen*, one of the four games in the *Thirty Years War* Quad. It may be remarked that this is one of the best of the set (another, the promising-looking *Freiburg*, should only be bought if an errata sheet accompanies it) and provides an excellent introduction to wargames for beginners as well as a pleasant and fast-moving game for old hands.

Artillery in the game cannot fire through units at other forces behind them, whether the intervening bodies are friendly or not (cannon balls are poor at distinguishing nationality), unless the guns are high up, like the Imperialist ones on the Schonfeld and Stoffelberg hills here. This is fine for the Imperialist player at the start of the game, though when he comes to grips with the main Swedish body in front of Hohlheim he will be blocking his guns' line of sight again, under the rules of the game. The main Swedish force, however, is balanced on the horns of a nasty dilemma.

1) It cannot retreat and abandon the guns, as these count for ten victory points each; the Imperialists could just wander over, seize the guns, and sit out the game.
2) If it advances to meet the enemy in the centre, then it
   (*a*) blocks the artillery
   (*b*) accelerates the main battle, which is a disadvantage, as the smaller clash in the south may well be won by the Swedes, who are tripled on their initial assault, and if this happens first, then there will be a big morale boost for the main battle.
3) If it stays put, cowering beside the guns, then the Imperialists marching over will be able to make the first assault, whereas otherwise the higher Swedish movement factors would ensure that they could do so.

It is important to note that the main Swedish force is outnumbered. This makes it clear that the Imperialists should attack, though they should send reinforcements to try and save the southern group on the Allbuch heights, with their valuable artillery. If the Imperialist player is lucky, his opponent will fall a victim to the 'Banzai' syndrome, and rush out to fight in the centre, with excellent initial results and eventual disaster.

The Swedish player should ignore the siren voices of his aggressive instincts, and sit tight doing absolutely *nothing* along most of his line except firing artillery barrages, and possibly sending cavalry south to reinforce the vital Allbuch assault, though the ride will be slowed by terrain. Only in the north should he send out an initially powerful force to take on the enemy cavalry which will try to penetrate north of Klein Erdlingen. The three 8-4 infantry units which the Swedes have on their left flank can be reinforced by a leader unit and a couple of strong cavalry detachments; this gives a strong local superiority, and they should be able to inflict severe casualties on the light German horse. Most important, there is little or no blocking of the Swedish guns, who will have an embarrassingly rich selection of infantry targets bearing down upon them, and who cannot fire through Klein Erdlingen anyway.

As the Imperialist infantry in the centre nears its goal, the 8-4s in the north can start to pull back, so as to be sure of being on hand for the desperate struggle about to ensue. As the enemy flings in its assault, the different unit functions suddenly become vital. Artillery can disrupt units, but not destroy them; disruption prevents attacking, halves defence factors, and slows movement to two; they are also vulnerable to cavalry charges. Disrupted units can be rallied if they are pulled back, especially if leaders are present. However, if casualties have been very high, the army becomes demoralized, and disruption units fall apart in confusion and terror. In *Nordlingen*, Swedish morale starts lower because it's clear that they are up against a hard struggle, so the Swedes need to inflict more casualties than their opponents, with partial demoralization setting in after the loss of 75 and 100 combat factors respectively.

As the Imperialists reach the Swedish lines, the defence splits up like this:

The *artillery* fires at point-blank range at the strongest units in the assault; two or three should be disrupted as a result.

The *cavalry* masses and charges the disrupted units, knocking them out and sending Imperialist morale plunging. Swedish cavalry (only) is doubled against disrupted infantry.

The *infantry* assault the undisrupted units, disrupting many of them, to lessen the next wave of attacks.

Against an indifferent Imperialist player, this should swing the battle decisively in the Scandinavian favour. However, the attackers can make the operation very difficult indeed, by employing the 'overloading' technique which we saw in the very different environment of *Panzerblitz*. Some things do not change with the centuries, and it remains a cardinal principle that defenders should always be given more than they can cope with at any one time, or not assaulted at all. If you look at the map, you will see that the Imperialists can reach the enemy lines at different times. The cavalry can do it in two turns, sections of the infantry can get there in three, while the rest (except two 9-3s on Schonfeld, who should go south) need four turns. To do it this way is suicidal mania. As each group arrives on the scene, they will be dispatched by the 3-pronged defence described above, in time to turn and deal with the next batch. Instead, the powerful attacking infantry should assemble within three hexes of the Swedish line, with a light cavalry screen in front to absorb the enemy barrage. When, and *only* when, the full force is ready, the cavalry moves aside and the full force of the infantry is sent into battle in a single awe-inspiring charge. The cavalry regroup behind.

Note that the Swedes cannot frustrate the plan by coming out to meet the first echelons of the attackers before the full Imperialist power is assembled, because this will inevitably lead to a pitched battle in the centre, which the Swedes are trying to avoid. Moreover, the drive north of Klein Erdlingen can be delayed until the main central assault is ready, since the cavalry in the north will soon be able to catch up if gaps appear. As the Swedish 8-4s will then be needed in the main struggle, this will make the going much easier.

The outcome of the game will depend on tactical expertise, on the success of the initial Swedish storming of the Allbuch, and on any strategic surprises which either side can bring off. If either succeeds in reinforcing the Allbuch battle just sufficiently to carry the day there, without heavily weakening the main force, then he is likely to carry the day, failing a brilliant tactical triumph on the main front by his opponent. The remarkable choice of alternative plans in such a relatively short (ten turn) game is what makes *Nordlingen* so interesting. The Imperialists probably have an edge, because they can reinforce Allbuch more easily, but the seeds of an upset Swedish win are certainly there if the Swede can outguess his opponent.

In summary, the main rules of thumb for games with units of varying capabilities are:

1) After fixing your strategic plan in general terms, decide which special units are most important, and hoard these for use when you need them.
2) If movement about the board is not extremely rapid for all units, commit a mixed force to each sector. A sector is defined here as the largest area such that all the special units in it can cover any part where they may be needed.
3) Take a position whose main strength lies in specialized units, assault it with more forces than those units can manage at one time; e.g. if the main strength lies in artillery, attack with more units than the guns can fire on, rather than piecemeal hoping for a lucky die-roll.

4) Examine each unit at the start, to decide in what kind of environment it will fight best, and try to ensure that it is committed to a sector where such circumstances are likely; e.g. an offensive aimed at a road should probably include fast transport, which may not be useful in the assault, but will be needed to use the captured road and make sense of the plan – if you wait till the road has fallen, you may find that the transport has become bogged down elsewhere.

8 WAVES AND SKIES

Much of the comment in the preceding chapters applies primarily to land games. Most wargames are in this category, for the obvious reason that most wars in history have been fought mainly on land. However, there are a number of air/sea games, as well as many games incorporating two or even all three elements.

The rules for air/sea combat tend to be very different in the games where this is peripheral to the main action from those where it is central to the game. In a game about the invasion of Normandy in 1944, for instance, the designer has a strong incentive to make the air combat relatively abstract, so that players can concentrate on the invasion rules which give the game its special interest. The abstraction of ground support by aircraft and naval guns is usually done in one of the following ways:

1) *Fixed bonus:* On each turn, one or both players receive a number of points of air or naval support which they can add to combat factors where convenient, within possible restrictions of range. Thus, a defending unit with seven defence factors might normally be safe from assault by more than twelve attack factors, giving an unsatisfactory 1–1. With two tactical air factors available, the odds could be brought up to 2–1. Because the points can be distributed individually in exact amounts to bring the odds up, a small ration of support points will have a disproportionately large effect.
2) *Simple units:* Alternatively, naval and air units may appear on the map, but in a simple way; they may be labelled as anonymous groups, or the combat system may be much simpler than on land. Thus, equal numbers of opposing aircraft may automatically eliminate each other, leaving any remainder for ground support.

GDW's *Avalanche*, on the Salerno landings, illustrates both systems.

The game is highly sophisticated in land combat, with one of the most advanced simulations of special unit functions appearing in any game yet published. In addition, it is at battalion/company level, instead of the more usual regimental level for this size of battle, so there are 1000 counters, and a full game takes a very long time to finish.

To avoid it becoming quite unmanageable, something had to be kept simple, and the designer reasonably settled on the ground support operations. Tactical air support follows the first system, with each side receiving a number of air factors each daytime turn to add to his ground combat factors. The amount available follows the historical flow of events in the air over the area, with the Allies becoming increasingly dominant once Montecorvino airfield has been captured and repaired by an engineer unit. Flak guns reduce the effect of air support, but apart from this, and the control of the airfield, there is no way to affect the balance of air power. You are the ground commander, and the air strengths are a parameter from outside which you have to live with and make use of as best you can. Naval support, however, played a rather stronger role in the landings, and is correspondingly simulated with the more advanced second method. A large naval force is positioned by the Allied player at varying distances from the shore, and the individual ships act as floating gun platforms. If the German artillery fires back, there is a simple rule to determine whether the ship sinks or not, and naval losses may also be incurred by an 'attrition' die roll representing ships sunk by mines or bombing.

Another use of air units is to interdict communications routes; this is an alternative to ground support in AH's *Anzio*, and appears on its own in SPI's *Panzergruppe Guderian* (the battle for Smolensk), in which three German air units per turn, as well as a more limited Soviet air and partisan effort, have the effect of braking rail and normal ground movement through affected hexes.

Such simple systems are easy to use optimally. The main consideration is placing any supporting air/sea units in positions where they can be available in as many skirmishes as possible. If it is possible to inflict disproportionate losses on the enemy ground support forces, this may be worth doing first, before committing yours to tactical help on the ground. In *Blitzkrieg*, for example, heavy strategic bombers can be used as an extremely effective weapon to interdict retreat routes (thereby effectively surrounding large front-line forces), but they can also bomb enemy cities, which act as air bases, and in the process destroy large enemy air concentrations, and force the remainder to base well behind the lines. It usually pays to do this first, as the air superiority thus gained can be used to interdict at will, without having to worry about the other side doing the same on their turn.

Games which focus on air/sea operations are totally different, though the total effect of the combats may be similar to that which a well-designed abstract rule would generate. Land combat rarely appears in such games, since it is difficult to represent realistically without a fair amount of detail, and will often have little effect on the short-term aerial and naval struggle. An interesting experiment is GDW's *Tsushima/Port Arthur* simulation of the Russo-Japanese war. The two games represent the naval and land wars respectively, and each has abstract rules to reflect the other theatre. Players interested in both can run the two games concurrently and eliminate the abstract rules.

There are many purely naval games, of which three appear in the top eleven in the SPI poll described in Part IV: *Frigate, Dreadnought* and *Wooden Ships and Iron Men* (which is also the most popular in AH's poll). Many players specialize in naval games, liking their mixture of subtle manœuvres and thunderous gunnery duels. Nearly all games of this type are at a tactical level, with simultaneous movement and varying degrees of damage on different parts of the vessels. A game where you could roll a couple of dice and sink the *Bismarck* on a double six would be boring and periodically farcical, though some allowance has to be made for sudden disasters like the *Hood.*

We will look at *Dreadnought* as typical of the species, though it should be emphasized that it is not as strongly tactical as some. *Wooden Ships and Iron Men*, originally designed by the tactical experts Battleline Publications, though revised and currently distributed by AH, goes into fire effects in far more detail, with gradual attrition of crews (weakening firepower and boarding parties) and structural destruction of both hull and sails (reducing manœuvrability and speed), and different types of ammunition suitable for alternative purposes. *Dreadnought* is at a less powerful magnification, so to speak, with less detailed fire effects but a correspondingly faster-moving game. Nevertheless, it features every twentieth century all big gun battleship built anywhere, and the tactical 'feel' remains impressive.

The map is made up of six blank blue hex-sheets; if a ship wants to run off the edge, a sheet not in use is transferred to that side of the map, giving an effectively infinite ocean. Each player operates a number of battleships, together with destroyer and cruiser screens, either in accordance with a historical scenario (e.g. Jutland or Surigao Strait) or after selecting them according to a points system. As in *Panzerblitz*, the latter is preferred by most players who have tried both and are not specifically anxious to simulate a particular battle, as the free-choice version adds an extra problem for the players and balances the game.

Once the task-forces sight each other, the players will attempt to manœuvre them to enable the strengths of their forces to come into play. The ideal is, of course, to have superior range as well as faster speed; one can then destroy the enemy at leisure at arm's length. The simultaneous movement system prevents this being automatic, however, since the underdog should be able to outguess his opponent part of the time, and either narrow the range to give his own guns a chance, or widen it in the hope of escaping altogether (a slender hope with inferior speed). With long range but slow speed the advantage cannot be maintained for long, and the faster vessel can dictate the course of the action. On the other hand, it will have to run a gauntlet of fire to close the range with the slower ship, and this may radically change the situation, affecting either firepower or speed.

Gunnery hits reduce firepower by half, or eliminate it if it has already been reduced; movement hits reduce speed by half, or eliminate it. To a certain extent, damage can be repaired during the action, so it is essential to move in groups: if one ship is damaged, it can be screened by the others while repairs are effected. An optional rule which gives the weak screening destroyers a major role is to allow these to lay smoke, through which no fire is allowed. It is often possible to route the destroyers so that not only the wounded vessels but the destroyers themselves are protected from enemy fire by the screen they have laid. This only buys a breathing space, since the enemy will probably close in at full speed to press home the advantage, but if the damage can be repaired before the screen is penetrated, then it may be possible to make the attacker wish he had restrained his zeal. This is particularly effective if the damaged ship is powerful but short-ranged; closing with it then becomes a hair-raising endeavour, but failing to do so allows it to recover full strength behind a screen. Illustration 16 gives a simple example; normally, of course, more ships and map sheets would be involved. In a hypothetical engagement between the wars, the *Arkansas* class American ship *Wyoming*, with a destroyer escort, has sighted the *Lutzow* class German *Graf Spee*. The four factors are: top left and bottom left: attack and defence; top right: range; bottom right: speed. The important thing to notice is that the *Graf Spee* is slightly outgunned, but has a substantially longer range, and faster speed. In the illustration its commander has succeeded in outguessing the American, and the range is seventeen hexes, which means that the German ship can fire but the *Wyoming* cannot. A lucky shot at this extreme range inflicts half damage on both gunnery and speed on the *Wyoming*.

This is a serious blow, since it reduces the speed to two, and tempor-

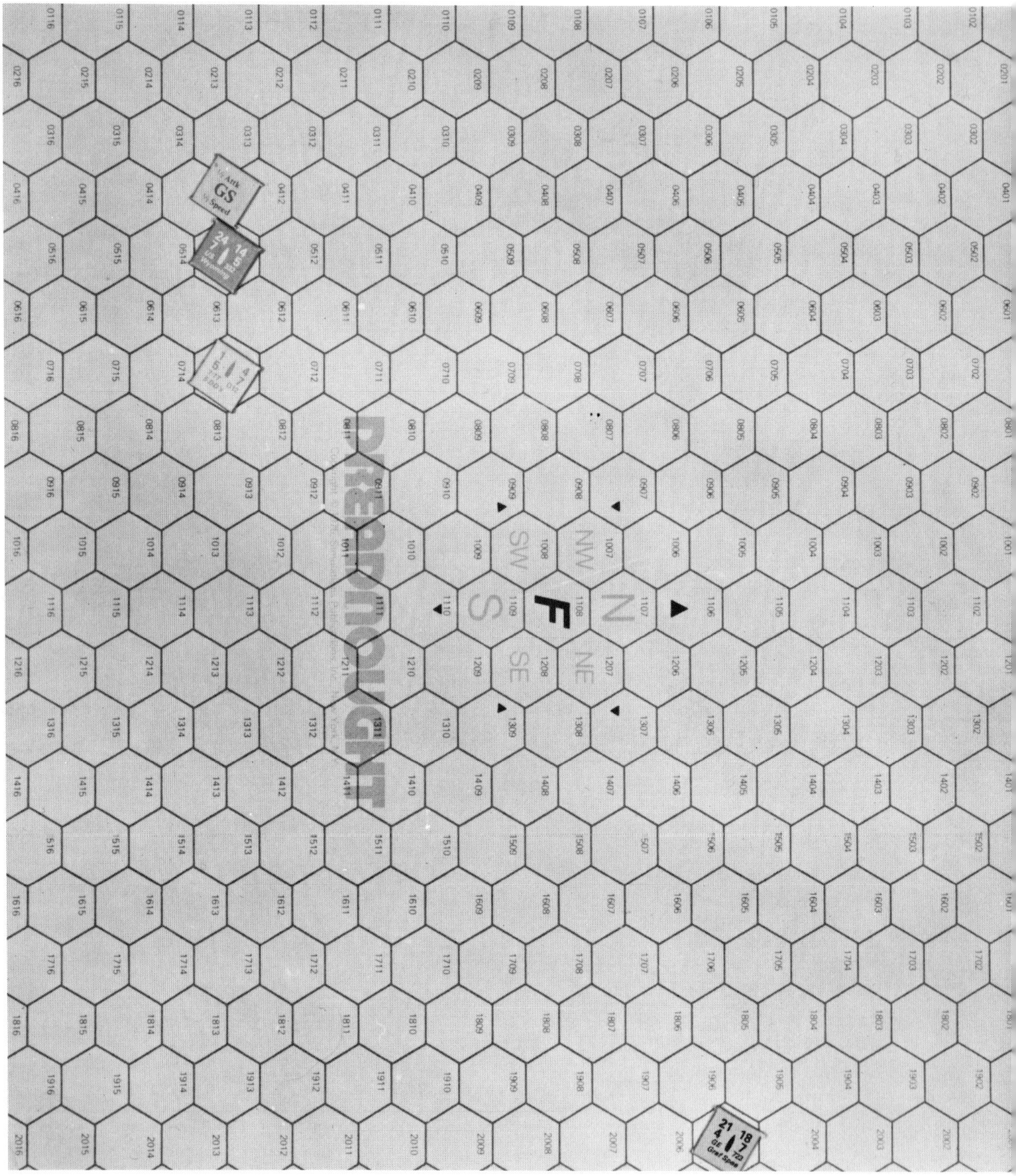

16 The *Graf Spee* draws blood.

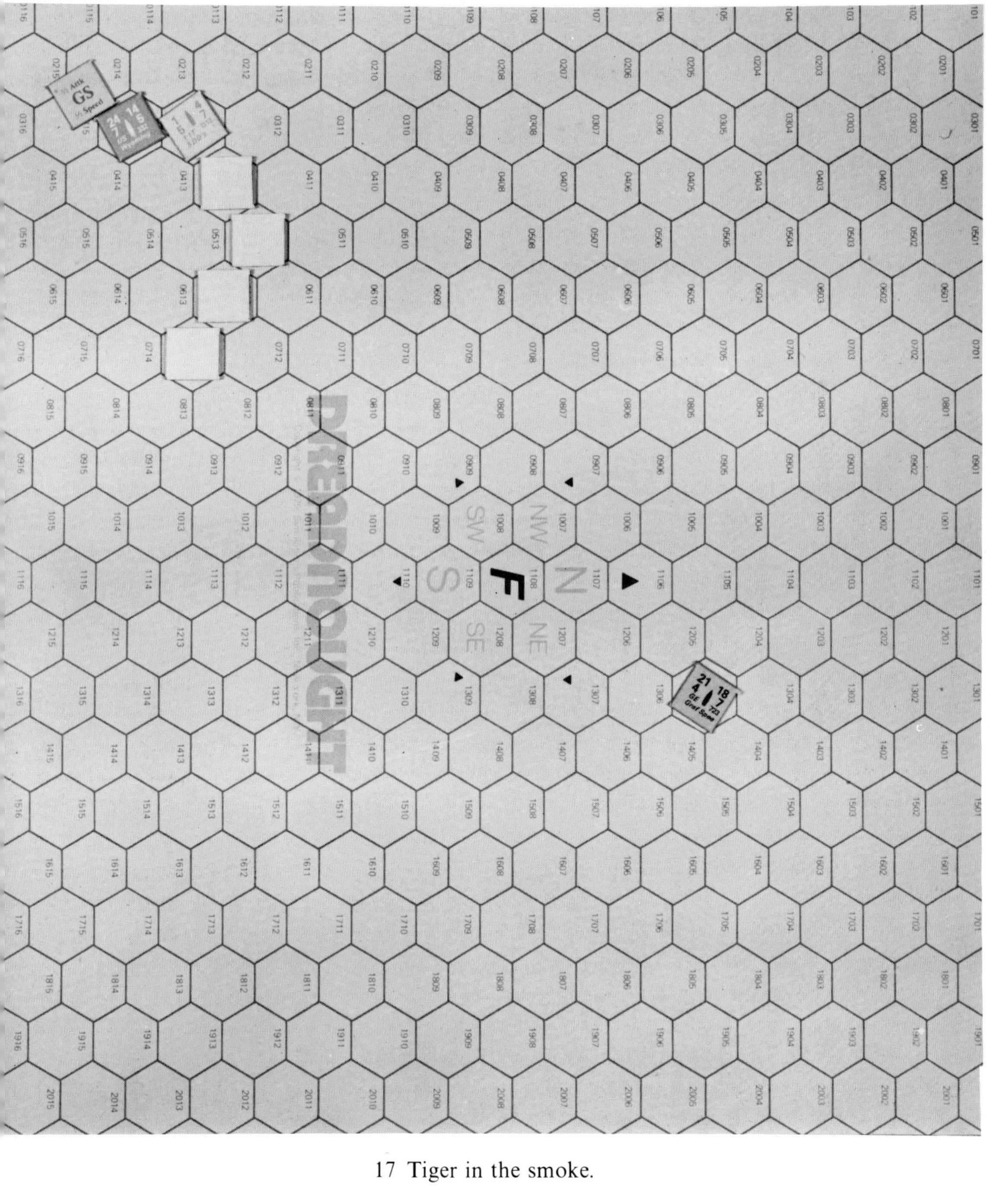

17 Tiger in the smoke.

arily makes the gunnery factor inferior as well. But the *Graf Spee* might have been better advised to sink the destroyer group, despite the superb opportunity to hit the main enemy without reply. Illustration 8b shows the position after the next movement phase. The *Graf Spee* has put on full speed to exploit its advantage before the damage can be repaired (an even chance per turn for gunnery; a one third chance for speed). The *Wyoming* has crawled southwest, away from its tormentor, while behind it, the screening destroyers have laid smoke (the blank counters) in between the antagonists, turning southwest as well, to protect themselves and lay the groundwork for a second smoke action next turn should the damage not yet be cleared on the *Wyoming*. By that time, the *Graf Spee* will be hard on their heels, and it will be difficult to lay yet more smoke thereafter in every direction from which the German might be firing on the following turn. However, two or three turns should see the *Wyoming* fully restored, and ready to use its superior firepower to cripple the enemy at the narrower range, before he can escape back to the fifteen-eighteen bracket.

Certain constraints restricted the moves shown, and are typical of naval games. No ship may increase its speed by more than 100%, or reduce it by more than 50%; guns do not fire straight ahead or behind, so the *Graf Spee* had to turn to give herself another broadside chance; to simulate the impossibility of sudden turns, dog-leg movement must have the second leg no shorter than the first – one cannot, for instance, steam six hexes southwest and then one hex northwest. This is why the German ship is still well to the north, and facing northwest. On the next turn, it can turn, steam three hexes southwest, and then four hexes south or west, depending on the commander's expectation of enemy movement. Because of this choice, the smoke screen may fail to cover the *Wyoming* a second time (the first smoke dissipates this turn).

This in turn illustrates the drawback of simultaneous movement from the point of view of skill. There is really no way to guess whether the *Graf Spee* will go west or south, or even forego its fire in the hope of closing right up and steam southwest all the way. It is all very well to represent this as an exciting duel of wits, but in fact it is simply luck (unless you believe in ESP). This is not necessarily unrealistic, since luck in guessing enemy plans (given alternative policies of apparently equal allure) must play its part in real life as well. Moreover, it *is* exciting, and not always quite an even choice; each player strives to interpret small scraps of evidence, in particular how their opponents have played on previous turns. And the alternative (consecutive movement) is decidedly unrealistic, unless the movement is something like one hex at a time (even then, simultaneous action is preferable for good simulation), though less chancy and easier to play.

Turning from pure naval games to pure air games, one finds striking similarities. Battleline's *Air Force* is one of the finest examples of an ultra-tactical game. Again, the map is blank, though the occasional land feature may be placed on it for the aircraft to attack. The emphasis is on the different performances of the machines, with each type having a sheet of specifications: speed, armament, ability to perform different manœuvres, vulnerability, and so on. The planes are placed on the map, but this only shows their location and direction; everything else, from angle (six alternatives) and height to degree of damage (which may be to half a dozen different parts of the aircraft). Movement is naturally simultaneous. There are a good many die-rolls whenever combat takes place, to determine hits and their exact effect, but in the long run the luck ought to even out, leaving skill as the decisive factor. Other air games use similar techniques, with variations: SPI's *Spitfire* and *Foxbat and Phantom* are simpler, AH's popular *Richthofen's War* has a particularly good variety of scenarios with air-ground attacks, and Lou Zocchi's *Basic Air Combat* is an individual-combat game using a sophisticated system of energy/manœuvrability accounting: planes can turn tightly or climb rapidly, but they are forced to lose height and briefly take it easy if they overdo their assault on gravity, to avoid stalling.

Pure air games of a different kind are the strategic campaigns: there have not been many of these historically, and the games are correspondingly few: *Their Finest Hour* and *Battle of Britain* deal with the RAF–Luftwaffe struggle in 1940, while *Luftwaffe* is an easily playable game on the bombing offensive against Germany, with more advanced and very much longer scenarios for the enthusiast.

Finally, there are the air/sea games, in which aircraft carriers play the decisive role: examples are *Midway*, *Solomons Campaign*, *Sixth Fleet*, *Coral Sea* and *Battle for Midway*. Since air attacks are frequently deadly, surface combat is rare in some games of this type (e.g. the Midway games), as the battle is decided by bombing raids. This gives the carriers their crucial role; sinking one not only advances the victory point total but strikes a hammer-blow at the vital enemy aircraft. Movement is almost invariably hidden in carrier games, with search procedures reminiscent of the children's game 'Battleships', except that the targets are on the move, and trying to avoid detection. Once contact has beeen made, the enemy must be kept in sight by reconnaissance aircraft while the friendly carrier fleet bears stealthily down upon its target: as soon as it is within range, the torpedo- and dive-bombers with their fighter escorts lash out. Defending fighters on combat air patrol and the flak barrage of the fleet try to hold the assault away from the carriers, and the defending forces' bombers in turn scramble into the air and try to trace the enemy back to their own ships.

Midway is a good representative of the type. The United States have a weakly-defended carrier fleet, plus an airbase on the island of Midway itself. Japan has a comparable carrier fleet, with massive surface reinforcements and a landing force coming into the game later, their objective being the capture of the island. The main US advantage is a better search capacity, and they have a chance to launch a raid on the Japanese carriers – if they can find them – at dusk on the first day, using night-time to slip away before retaliation finds them. If this has weakened the Japanese sufficiently, it may be possible to sink the landing force when it appears, or alternatively to wreck the rest of the carriers. Failing this, the odds are on Japanese victory, since the combined air and flak defence power of the full Imperial fleet is almost overwhelming.

Midway uses the battle board technique also seen in some purely naval games: when an air (or surface) attack is launched, the defending fleet position is set up on a separate board, for the resolution of the tactical combat.

The tactical system used varies greatly from game to game, but the *Midway* system seems as good as any so far, and makes a good concluding example for the tactical part of this book. The rules are these:

1) The defending ships must set up on the battle board at least one space apart from each other. They can all direct their anti-aircraft fire two squares (straight or diagonally) or less away, but each ship can direct its AA guns only against a single square. The AA factor is the right-hand one on each counter. The American AA factors are all three, except for the *Atlanta*, with six.

2) The attacking aircraft are divided into dive- and torpedo-bombers, written with the number of squadrons involved as D and T respectively: thus, a T4 represents four torpedo squadrons. Torpedo-bombers are placed on the square on either side of a ship under attack (optional rules on anvil attacks on the bow, and wave attacks, will be ignored here for the sake of simplicity, and it is assumed that fighter combat has been resolved already). Dive-bombers are placed *on* the ship. This means that there will be cases where a ship can defend a colleague from dive bombers but be out of range of a torpedo group, or vice versa.

3) Illustration 18 shows about half the American fleet, set up in what

is probably the best position to repel an attack on the carriers which is dominated by torpedo bombers. The vital AA ship *Atlanta* is covering both flanks of both the carriers in the middle of the fleet, and two of the other three escorts are also able to reach each flank. Had the main attack force been thought to be dive-bombers, it would have paid off better to have a box-like formation of three rows of two ships, with the carriers still in the middle, since this would have given maximum anti-dive-bomber cover to both of them. However, this would have left the outside carrier flanks exposed, especially on the side away from the *Atlanta*, so the formation is unsuitable against torpedo planes.
4) Combat is decided by comparing the number of squadrons on any one square attacking one ship with the amount of flak firing at that group; if there is no flak, the defence factor is set at one. Thus, if a T6 attacks the *Yorktown*, and they shoot back without support from the other ships, the odds are 6–3, or 2–1; if the *Yorktown* gunners are firing instead at dive-bombers on the *Enterprise*, then the *Yorktown* is attacked at 6–1.
5) The CRT can be summarized like this:

ODDS	NUMBER OF HITS
1–1	1 or 2
2–1	1, 2 or 3
3–1	2, 3 or 4
4–1	3, 4, 5 or 6

The alternative outcomes are more or less equally probable. Attacks at under 1–1 are unlikely to succeed, while attacks at over 4–1 automatically sink the defending ship. Losses are taken in attacking aircraft in almost all cases, but need not concern us: if the attack can succeed in sinking one or both carriers, that will be worth piles of losses.
6) The carriers take five hits to sink, as do the escorts except the *Atlanta*, which takes only three. The objective of the attack is to sink the carriers, or, failing that, to make sure that they can be sunk on a future attack. The attackers have T32 and D13, which can be divided as shown.

Napoleon Clausewitz Smith straightforwardly sets up the attack shown in the illustration. (Where more than one aircraft counter on a square is shown attacking the same ship, this is because no sufficiently large single air counter was available; the planes are still counted together in one attack.) He has noted that if he divides his force into two, with T18 and D7 attacks on one carrier and T14 and D6 attacks on the other, then he is unlikely to sink either enemy carrier, with the defensive flak sending up six factors against each attack except

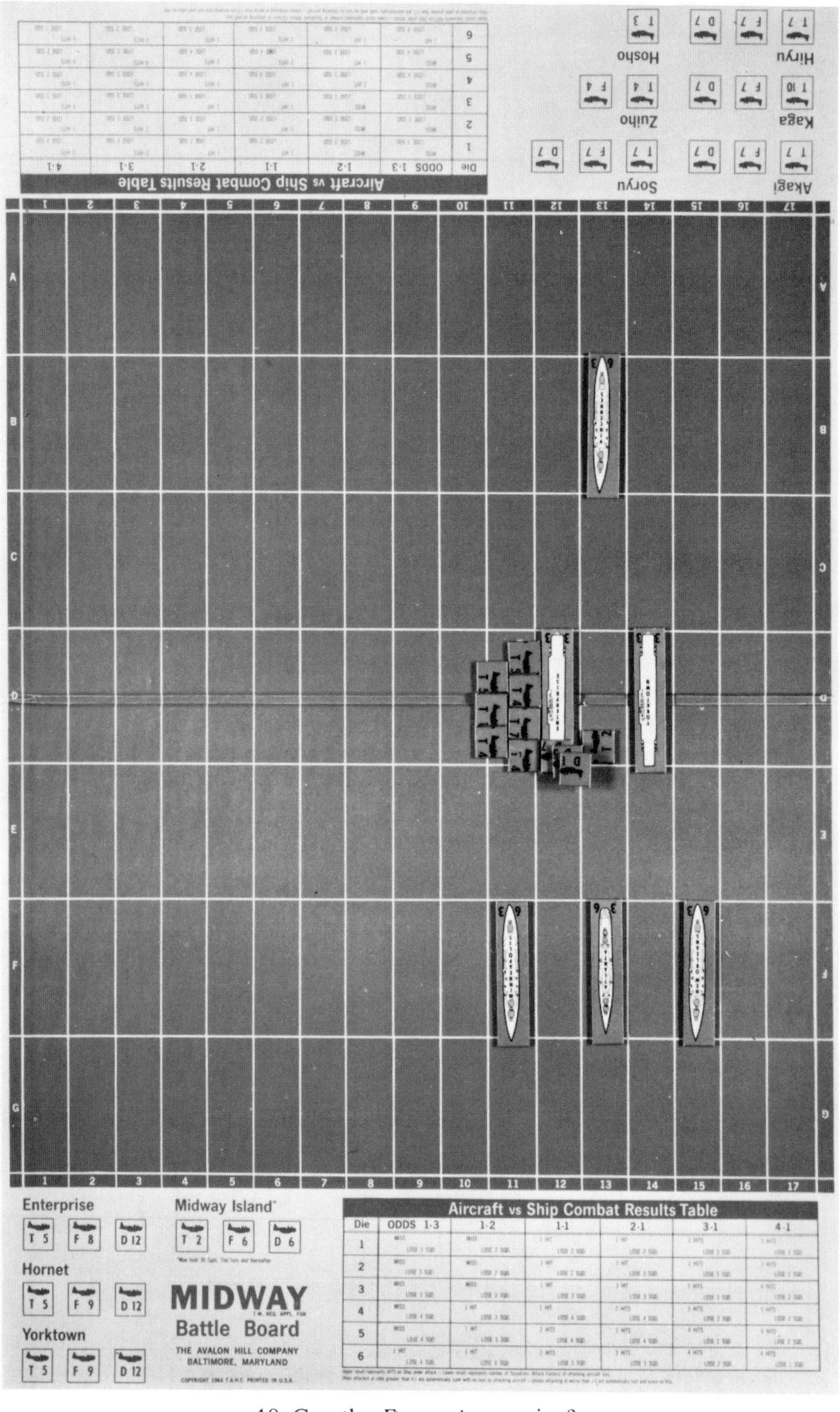

18 Can the *Enterprise* survive?

the last, against which only three can be mustered. This would give a 3–1 and 1–1 on one carrier and a pair of 2–1s on the other, gaining about four hits on each (going by the summarized CRT above). This might mean that they were both sure to sink next time, except for the fact that it leaves them both afloat, with their aircraft – and with the greater US search capability, it is possible that the US will get in the first attack after this, and sink the Japanese carriers – in which case there will be no second chance, or only with such a weak attacking air group that the American screen units will dispose of them.

So Smith opts for a mass attack on the *Enterprise*, with T30 attacking the starboard, a diversionary T2 attacking the port, and D13 diving in from above.

But Smith has made a serious error; this plan is actually considerably inferior to the other alternative. Moreover, there is a third plan which should sink one carrier, and give an excellent chance of destroying the second on another raid.

Problem: (*a*) Find the defensive flak fire which stops Smith's plan from working.
(*b*) Find the optimum attack.

Use the summarized CRT, assuming the outcomes shown to be equally probable. Remember that ships cannot divide their fire. The *Atlanta* and the *Enterprise* could fire at the same group of T30 in the illustration, giving odds of 30–9 = 3–1, but the *Enterprise* could not strip off one of its flak factors (leaving a 30–8, still 3–1) to use for firing at other aircraft.

PART III

WINNING

It is a feature of complex games that playing them holds most of the enjoyment, and actually winning is just a pleasant way to round them off. In a simple children's pastime like Ludo or Snap, it is impossible to be surprised, refreshed, and delighted by the mechanics of the game, and all that is left is an occasionally ferocious zest to get better die-rolls or cards than the other players. Most chess, bridge or Go players, however, will enjoy brilliant play, even if their relish is somewhat muted by the fact that it is being practised by their opponents on them. The same is true of wargames: you may feel a little sick at losing Berlin in 1942, but it is hard to resist a horrified thrill at that masterly armoured pincer which made it possible – and there's always a next time when you'll find the definitive counter-strategy!

Nevertheless, most players would rather win than lose, and in the next two chapters you will find suggestions that should help you get an edge against otherwise equal opposition.

Chapter 9 looks at the technical details for gaining tactical advantage.

Chapter 10 draws together the strands of the book to give a rounded picture of a good player – and how he got that way.

9 THE SHORT CUTS

Wargames have a number of short cuts which make success easier and give an edge against opponents not aware of them, however expert they may be in other respects. The rules of thumb in this chapter are some of the more important ones: others will occur to you as you develop your individual style of play.

PUT THE RULES UNDER A MICROSCOPE

Extraordinarily wide misconceptions arise over what the rules to a game actually say. A player reads what he thinks makes sense instead of what the designer intended, he writes an article on good play based on the mistake, and 90% of the readers allow the error to enter their own assumptions. Always read the rules carefully and note what they really say and what they don't mention.

For instance, in *La Bataille de la Moskowa*'s small scenarios, only part of the armies on the Borodino battlefield are used, and rear areas of the map are empty. The rules state that replacements must be brought on a large number of hexes away from the nearest enemy unit. Very well: send a cavalry unit round the flank into the enemy rear area, and he will be forced to bring reinforcements so far back that they will be totally useless. Or you may prefer to agree with your opponent to outlaw this tactic in the interests of realism; other ways to slip through gaps in the rules may be less absurd but equally useful. Avalon Hill gave an entertaining look behind the scenes on this subject in an article in *The General*. The designer of *Tobruk* was an expert on military affairs rather than a long-standing wargamer, and he prepared his first draft of rules with allowance for common sense. AH explained to him that it's no good assuming that all wargamers are willing to use their common sense in rule interpretation – if the rules don't say that the main Russian army not shown in the scenario cannot stop a French cavalry regiment riding into their midst, then they can't! While this may seem unreasonable, there is little alternative in general to going strictly by the rules, since there are many cases where the players' ideas of what is common sense will be honestly different.

Knowing what the rules say is important, but one does not have to be a walking compendium of them, and especially when playing a game for the first time, it is sensible to settle for knowing the general import of them, and looking up fine points where necessary. As one gets to know the game, the rules will stick in one's mind, and the important small points which can change the course of the game will become noticeable.

Above all, play according to the victory conditions, and not on some general principle of killing a lot of enemy units or pushing him back. This mistake may sound obvious, but a great many players make it even in games they know well; they know that they are supposed to be capturing Kharkov, but the temptation to fight out a battle somewhere else with optimum force pushes the main priority to the back of their minds.

WORK OUT THE EXACT ODDS

Few players with any experience fall into the trap of piling 'a lot' of units into an attack without counting the strengths, only to find that the odds are 29–10, rounding down to 2–1 with 9 factors wasted. What is more common is underuse of strong units. Suppose you need to get 30 factors to achieve reasonable certainty of furthering your general aims in a particular battle, and you have four 9-factor units among the forces available. It is then easy to say 'Fine! $9 \times 4 = 36$; that's enough; so much for that sector.' Fifteen hexes down the line, you find you are short of five factors, but by then have forgotten that you had more than you needed in the first battle.

There are two linked rules on this: always use the minimum force to achieve what you want, unless there is absolutely nothing else to do with the surplus, and always try to do something with every unit (even if it is just consciously placing it in a useful reserve position). Units which are only within reach of one sector should be moved first, with the high-movement-factor and well-placed forces which can reach several areas of combat moved last when you can see where they are needed.

It is not always sufficient to commit the minimum force to obtain favourable odds. You may want to hold the position with a strong group after it has been captured by your attack. Possible casualties must be taken into account; games with 'exchange' CRT results can result in the elimination of all the combatants at 1–1, leaving the battle won but the field empty and ripe for seizure by the enemy on his turn. Another case is the games with unknown unit strengths, either because of hidden movement or because some forces are 'untested' and do not have their strengths displayed to either side before they enter their first battle (*Panzergruppe Guderian* and *Invasion: America* use this technique, designed to avoid exact calculations of combat odds).

KNOW YOUR OPPONENT

Most of us get into the habit of playing with the same people. This has its drawbacks, as what works against aggressive Smith may lead to disaster against cautious Jones. There is a natural tendency for the regular opponent of Smith to turn into what a physicist might call anti-Smith: the antithesis of Smith, filled with plans and stratagems which only succeed because Smith is unable to change his style to meet them. If this is done *intentionally*, however, it makes excellent sense. If you are sufficiently flexible to change your approach according to your opponent, then you will have a lasting advantage (until they do the same for you). If they like risky attacks, gratify their suicidal impulses by making plenty of opportunities available. If they enjoy a slow, careful build-up, then this tells you something about their likely strategy and production in economic games.

This sort of thing can be taken too far, until it becomes gamesmanship, putting winning before enjoyment, which will spoil the fun for both sides. Within reason, however, it will lead to the sort of game which each player enjoys, and improve the play of both. Thus, if you think that your opponent takes excessively daring chances and set out to encourage this, the outcome will be first that he has a delightful time indulging his alleged weakness, and secondly that it will either gradually dawn on him that he is taking too many risks ... or it will dawn on you that his approach works after all!

GAMBITS AND RISK-TAKING

Calculated risk-taking and recklessness are quite different. In some ways, risks are the essence of wargaming. Given time and patience, it is theoretically possible for a chess or Go player to work out the exact consequences of his actions, making risk-taking impossible except in the sense of gambling on an opponent doing (or failing to do) something. Not so in wargames, as we have seen earlier. One can never be quite sure how things will turn out. It has been argued in this book that a good player will be ready for any possible outcome. While this is completely true in the sense that no good player should ever be caught saying, 'Good God! How did *that* happen?', it does not follow that one should never take risks. The important thing is to do it with both eyes open.

What kind of gamble is worthwhile? There are a few general rules on this.

1) Take gambles when ordinary methods fail. In *D-Day*, using the usual rules, the Allies seem to have an edge, so long as they invade in a lightly defended area a long way from the Rhine. If they choose a heavily defended northern area, they may have a chance of winning swiftly and dramatically. If you are in a hurry or want instant drama, fine, but if you want to play the Allies as well as you can, then you should save the daring tactics for when and if the cautious approach bogs down.
2) Conversely, if ordinary methods seem likely to fail, then be more receptive to exotic ones. It is particularly useful here to know the game bias, if any. We saw *Winter War* in Chapter 3: it is thought by many to be a very difficult game for the Russians to win. If the Russian player in a particular game agrees, then he should not be content with proving it by an orthodox offensive, but fling his units into low-odds attacks against the key fortification lines. If they mostly succeed, he will probably win. If (more likely on the whole) they mostly fail he will lose, but then he expected to lose anyway.
3) Look closely at the strategic effects of a possible gamble. An attack which risks 11 factors for the possible gain of 3 is superficially unattractive, but if they have special functions, or are holding a major position, then it may make sense. Try to quantify the gains. Care is needed here: if gaining the position would lose the enemy five replacements a turn for the remaining six turns of the game, it is *not* worth thirty factors: the replacements arriving later are worth less than units available now, and it may not be the last chance to capture the position. This sort of calculation leads on to the final point.

QUANTIFYING PROBABLE CASUALTIES

Suppose that we are considering an attack on a doubled 5-factor unit in a 'classic' game, and it will require two soak-offs at 1–3 with 6-factor attackers, in addition to the 30 factors needed to make the main assault at 3–1. Is it worth it? The answer hinges on semi-unquantifiable factors like the value of the doubled position in strategic terms, but it is useful to know the expected losses on each side, as one aspect of the problem. Looking up the CRT (shown in Chapter 2), one finds that a 3–1 gives a 1/3 chance of eliminating the defending 5 factors outright, a 1/3 chance of eliminating them in an exchange with 10 attacking factors, and a 1/3 chance of merely forcing them to retreat. Giving each of these losses the weight of the probability that they will happen, we find that the expected casualties average out at 5/3 plus 5/3 plus 0 = 10/3 for the defence, and 0 plus 10/3 plus 0 = 10/3 for the attack. The soak-offs both have a 1/3 chance of resulting in attacker elimination, and a 2/3 chance of merely making them retreat. The expected losses are therefore 6/3 plus 0 = 2 in each case. The net result of the combat will therefore be on average four factors more lost by the attack than the defence. Of course, the capture of the position may be worth it, or, alternatively, some of the results might lead to major consequences (e.g. the loss of the two soak-off units might allow the main force to be surrounded).

If there is a choice of different attacks with different factors involved, this method is quite effective at showing the best policy. Another example using the 'classic' CRT is the observation that a 1–4 soak-off against 24 factors is cheaper than the other possible attacks, with an expected loss of 3 factors of the 6 needed, against between 3 and 4 factors if attacks at 1–3, 1–5, or 1–6 are made. Since one will often be making soak-offs, it is worth knowing that in the long run a policy of always using the smallest force (at 1–6) does not pay off.

This method is too time-consuming for normal purposes, but can be used for particularly crucial decisions, or for postal play.

CRTs in modern games often vary somewhat from the classic pattern. Sometimes only the defender incurs losses (and is then allowed to counter-attack), and sometimes the difference in unit strengths rather than the ratio is used (so a 21–10 is not a 2–1 but a plus 11); this is for the player's convenience and reflects logarithmic adjustments to the combat factors. In all cases, the basic technique remains the same.

10 THE BROAD SWEEP

Tactics and strategy have been dealt with separately in this book, because they have distinctive elements which are most easily learned in isolation. But it is, of course, no good expecting to be able to keep this up in actual games. If you are contemplating storming a bridge because it will impede the enemy flow of reinforcements and enable you to isolate a big hostile force from its brethren over the river, then you have strategic as well as tactical considerations, plus certain aspects with some features of both. If you try to decide on strategy first and tactics later (or, still worse, *vice versa*), you may end up with an unsatisfactory result. What is needed is a subtle, slippery way of thinking, which interweaves strategic and tactical planning and uses the techniques of each automatically when required. In the case above, a good player might think along these lines:

1) I am worried by those reinforcements. If they keep up they will disrupt my offensive and force me to send reserves to the weak eastern sector. What can I do about it?
2) Maybe I could capture the bridge?
3) What's holding the bridge? What are the chances of forcing it off? A 'defender retreat' result would do nicely; I don't need to inflict heavy casualties.
4) Suppose I did take it, what other strategic effects would it have? It would isolate that big enemy force, and they are too tied up at the main front to be able to get back and dislodge my troops.
5) Can I afford the tactical shift of forces from my west front to the bridge? They will have to stay there for some time, with a group at the bridge and another guarding the flanks. I don't want to actually blow the bridge if I can help it, as I can use it myself later.
6) It looks good – I'll do it! Or, wait a minute, wouldn't this alternative plan be even more effective?

And so it goes on. This kind of twisted mind evolves with practice, and gradually the process becomes second nature. The last step, however, is an important one which even good players frequently forget: however excellent an opportunity, always look to make sure there is not something better – not so much out of perfectionism, but because your opponent may well have seen the obvious chance and have something prepared to counter it. If there is a subtler opportunity which you can exploit, then the chances are that he will have overlooked it, as when he was planning his move, he did not know the effects of combat on that turn, so he was working with incomplete information.

The fact that it is impossible to look ahead with complete accuracy should absolutely not prevent you from doing everything possible in this direction. Both casualty rates and positional advances and retreats should be 'paced', and if they are running at an unfavourable rate then you should do your best to break the pattern before it is too late. If you have bad luck with a series of individual die-rolls, do not absent-mindedly fall into the statistical heresy of saying 'my luck is bound to turn soon' – Lady Luck does not adhere to human concepts of fairness.

When in doubt, hand your opponent a copy of *Gone With the Wind* or some similarly lengthy tome and tell him to go away and read it, while you occupy his chair and look at the position from *his* angle. It's odd how different it looks. Positions which you casually assumed were impregnable to his forces suddenly sprout nooks and handholds for an attacking force. Sneaky uses of rail lines spring to mind. You suddenly realize why he put those cavalry forces in that wood. In theory all this should have been visible before, but you have changed your standpoint: you are no longer setting up defences but looking for ways to break them down. If you use your own ingenuity against yourself before your opponent has had a chance to use *his*, a great deal of trouble can be avoided.

The more you play any wargame, and in particular the more you play the same one, the greater will be your instinct for the positional questions which arise. An experienced *Sixth Fleet* player will be able to tell almost at a glance if the Nato forces have lost the battle for the Aegean after a few turns, or whether the carriers from the Western Mediterranean are going to arrive in time to turn the tide. A further look will show him the weak units on each side and the general trend which the next player-turn should follow. He can then, if he wishes, settle down to work out the exact tactical dispositions, striving to exact every percentage point of a chance to sway the struggle his way. If he is really bent on finding the best possible move, you can safely go into town, eat a three-course meal, and walk home. There he will be, still planning away ...

Few people will want to play face-to-face like this (though I have sometimes spent an afternoon on a crucial move in a postal game), so experts and beginners alike have to cut back on something. The difference is that the expert will be able to map out the general strategy at lightning speed and then reach a pretty fair tactical implementation rapidly, while the beginner will find that he does not really have time to do more than choose a relatively obvious strategy and the simplest way to approach it tactically. There is a lot to be said for using chess clocks to reduce game length, and avoid excessively precise calculation of exact odds, which is in any case unrealistic. Wargaming can be treated as an art or a science; arguably, it is most enjoyable if neither player strives for mathematical perfection.

EXPERTISE

Practice is one element of becoming an 'expert' at a game, but on its own it does not seem to work miracles, especially if the practice is solo or always against the same opponent. There is a computer technique called 'bootstrapping', based on the idea of pulling oneself up by one's bootstraps. This is typical computer illogic, as anyone who has tried will agree that the literal process can't be done. In the same way, it is difficult to learn to play well without encountering a number of good players. I should not care to repeat my first game of *Panzerblitz* with another player: all my carefully-prepared (in solitaire play), allegedly brilliant techniques simply did not have occasion for use, as the game went quite differently to what had always seemed to me natural for *Panzerblitz*.

There are two ways of getting this experience, which is worth doing even if one doesn't care about becoming a great player, because it adds interest and variety to the games. I am exempting people who happen in any case to be surrounded by active wargamers on whom to cut their teeth. One can subscribe to one of the magazines on the games giving hints on actual play, discussions of strategy and blow-by-blow descriptions of actual games, often complete with a referee loftily pointing out where the players went wrong. SPI's *Moves* is aimed at this area, and it features heavily in Avalon Hill's *The General.* Several other magazines, both professional and amateur, also specialize in this field (see Appendix B). The drawback about some of these is that they deal with a lot of games, so that the beginner will frequently be tantalized by detailed discussions of delectable-sounding games he has never seen. Players with a fair number of the recent games of a company will, conversely, find each of their range dealt with in turn, and the magazines are ideal for these players. The amateur publications also contain reviews of many new games, so they can be useful as a means of keeping abreast of developments and selecting the most interesting recent products.

The second way to gain experience outside one's immediate circle is to play by post. Because there is more time for each move (though less need to keep aside five hours at a time, as in the average face-to-face game), and because opponents of every strength can be encountered, this improves playing ability with extraordinary rapidity. The drawbacks are that the game will take a long time to finish – a year or more is not uncommon – and either it must be kept set up where neither beloved relatives nor equally beloved pets can get at it, or it must be set up afresh each turn. In many ways, though, postal play is more fun than face-to-face. You can choose your own time and place to work out your move, you are never hurried by an impatient opponent thinking of the last bus home, or bored by a player of the let-me-just-check-this-once-more type, and you become part of an international fraternity of players. Wherever you travel, there is a fair chance that your club can tell you of members in your area, possibly people you have played by post; they will often offer you a bed for the night, probably on condition that you don't use it and spend the time playing *Terrible Swift Sword* instead.

During the early years of the hobby, there were several large clubs in the USA devoted to wargaming, but it is now thought that they placed too much emphasis on organization and intensive competition, and they have gradually faded from the scene, although smaller, local-based, groups continue to exist. The only club to survive intact has been AHIKS (the Avalon Hill Intercontinental Kriegspiel Society), whose illustrious title belies its quiet and unassuming objectives: friendly contacts between mature players not interested in cut-throat competition. AHIKS have a minimum age limit (occasionally waived) of 21, in the belief that this will increase the likelihood of players being willing to continue games to the end even when they start to lose, and not 'drop out' of the game without warning to their opponents. My impression is that 'drop-outs' come in all shapes and sizes, regardless of age, but it is not unlikely that AHIKS's serious-minded image does attract a high proportion of responsible players.

The main alternative to AHIKS is probably the National Games Club, a British-founded organization with sections for Diplomacy, Scrabble and chess as well as wargames. I founded the wargames section in 1973, and am at present General Secretary of the club. The

NGC is increasingly international in its membership and activities, and attempts to maintain a lively and relaxed atmosphere, with especial encouragement to members with new ideas and game projects. Instead of an age limit, the drop-out question is tackled with a system of game deposits, which are larger than the game fee: players receive these back after a completed game, while victims of drop-out opponents get their deposits as compensation. The idea is to guarantee a satisfactory game or one's money back.

A third, more recent, group, is the Conflict Simulation Society, noted for their magazine *Outposts*. The addresses of the three groups are in Appendix B.

The organizations tend to have a 'drop-out' level well below that on the 'open market', as exemplified by 'opponents wanted' advertisements in magazines. It has been estimated that up to 80% of games started in this way end in mid-session as one side starts to lose. Although it is possible to find reliable opponents in this way, saving club membership fees, one must be prepared to have a series of unsatisfactory finishes before a reliable opponent is found.

Not every game can be conveniently played by post, and some are actually impossible for this purpose, because of the need for constant interaction between players during a move. In most cases, however, there is little problem. Avalon Hill sell play-by-mail kits for many of their games, but these are rather expensive and not essential. The simplest way to send positions is to maintain a record on a sheet with one column for every move, and one row for every unit, with the latest positions written down the relevant column for the latest turn in the intersections. Die rolls can be resolved by using the sales-in-hundreds figures in American stock market reports (one can take the last digit for a pre-arranged stock's sales, with 7, 8, 9, and 0 becoming 1, 2, 5 and 6 respectively or use a ten-digit CRT which is sometimes available in the game or a play-by-mail kit). The clubs have their own simple systems, the game organizer using his neutral position to replace the slightly laborious stock market approach. Alternatively, you can just trust your opponents. Combat, with advances and retreats pre-ordered for the range of possible results, is written on a separate sheet.

CONCLUSION

Like most recreations, wargames are more interesting if both sides are good players than if they are hopefully pushing the pieces around and concentrating on mesmerizing the dice. If you could make a reasonable attempt at most of the problems in this book, then you should now be able to meet most players with some confidence. You will never win all your games, but few people would enjoy anything they invariably won! Some games will have a built-in bias, others can occasionally turn against the best players as they run into a steady stream of unfavourable die rolls – but whether you are winning or losing, whether the luck is with you or against you, you will be presented right until the end with the constant choices of strategic plans, tactical devices, and sudden gambles which make wargames fascinating, enticing and richly enjoyable. If this book has given the reader a taste of these qualities, and encouraged him to try the hobby for himself, or deepen his interest in it, then it has succeeded in its task.

PART IV

Simulating History on $10 a day

A conducted tour of wargames in print

The following is a list of every professionally produced wargame which I know to be available when this book goes to press, plus some scheduled releases in 1977, with notes to help the reader decide which will interest him most. These notes are drawn from my own experience and that of friends, together with the official views of the producers, and in no case should be taken as authoritative, official, or anything but honest personal opinion. However, I have seen a lot of games over the last decade, and think my assessments fair and balanced. Knowing the pitfalls of personal prejudices, I have generally tried to avoid the 'avoid this horrible rubbish' or 'rush out and buy it at once' sort of comment, and concentrate on showing the strong and weak sides of each game.

To buy one of the games, or get a list of the latest products, write to the company, who will send you an up-to-date catalogue and price list (I have not included prices, as these change rapidly). If you have a local store which stocks the game, or a local wargames club, this is better still, as then you can have a look for yourself.

As I am not yet a centenarian, I have not had the time and opportunity to try out every game on the market, and some of the descriptions are briefer than I should like, and in a few cases I know nothing whatever about a game except its name. Comments on any games, whether or not they are already on the list, would be welcomed by the author, for possible inclusion in a second edition.

The big companies, SPI and AH, have regular reader surveys in *Strategy and Tactics* and *The General* to discover the popularity of different games. The SPI survey covers nearly all games, while the AH one deals only with AH games. It is to be expected that the SPI respondents will tend to be particularly keen on SPI games, as they read the magazine which includes 6 such games every year; on the other hand, they may be less favourable to these than people who have singled out the games for separate order. Conversely, the AH poll is likely to be friendly to AH games; it should also be noted that the AH response is more bunched around the average rating than the SPI one, probably due to slightly different phrasing of the questions. It is also very important to remember that the polls are taken from hard-core players, so that complex and innovative games get a particularly warm response and *vice versa*; thus the SPI table is topped by *Drang Nach Osten* and *La Bataille de la Moskowa*, two of the most complex games around, while the very elementary *Kriegspiel* comes a poor last in both polls. There are 202 games in the SPI poll; 25 in the AH one.

An entry like this:

AFRIKA KORPS, AH (SPI 5.6, 150/AH 6.10, 14)

means that Afrika Korps is made by AH; it got 5·6 points on the SPI poll's 'acceptability' scale ranging from a perfect 9 to a repulsive 1; it came 150th of the 202 games in this poll; it scored 6·10 on the AH poll's cumulative scale; it came 14th of the 25 AH games. Just to be annoying, the two companies both have 1–9 scales, running in the *opposite* direction, so I have taken the complement of the AH ratings so as to make both run up to 9 as the best score.

The SPI poll is taken from *Strategy and Tactics* 57. The AH poll is taken from *The General* Vol. 13, no. 2. Games with identical ratings have been grouped in the average position; thus games in the SPI poll with 5·8 occupy the places from 133 to 140, and are all noted as 136th (rounding upwards).

AH and Battleline games are all boxed; SPI games except folio type usually likewise; DCC-AWA normally use a snap-lock plastic case; most other companies tend to use plastic pouches.

A final note: don't put too much into a small rating difference, especially when a small poll sample is indicated for that game. A difference of half a point seems to reflect a genuine variation in popularity; smaller gaps may not be that significant, as fashions change and new games in particular tend to fade a little by comparison with exotic newer products as time goes by.

Charles Vasey has kindly supplemented my personal knowledge with notes on a number of games which I have not seen; these are denoted (cv). Games denoted (cv-np) contain remarks from both of us. The *Island War* reviews were first published in *Battleground* and are by Marcus Watney. I am also obliged to Walter Luc Haas for allowing me to use *Europa* for source material and I have drawn on Richard Berg's review column in *Moves*.

AFRICAN CAMPAIGN, Jedko (SPI 5·4, 169) Brisk North African Second World War simulation.

AFRIKA KORPS, AH (SPI 5·6, 150/AH 6·10, 14) Operational 'classic' game of the North African war. Extremely mobile, with no fixed lines lasting for long once probes sweep round the flanks. Well-balanced, fast-moving easy to learn, but rather luck-dependent. Length 2–4 hours. *See Panzer Armee Afrika* and *Rommel* for good alternatives.

AFTER THE HOLOCAUST, SPI (not in polls) Basically a four-player game, with scenarios for one, two or five instead, on the rebuilding of America after a nuclear holocaust. The country is split into four regional governments, and the emphasis is on economic aspects, with a fascinating choice of strategies: investment in agriculture, mining, fuel, manufacturing; taxation; levels of government spending; guns *v.* butter; trade; bank lending. The areas can cooperate, in which case the one with the highest social state (popular prosperity) will win, or they can attempt to dominate each other militarily, taking scarce resources from the economy. Highly innovative.

AIRBORNE, Jagdpanther (not in polls) Fifteen scenarios on every kind of paratroop action (Crete, Bruneval, Arnhem and Indochina appear). Each turn is three minutes, each unit a single vehicle, ten men, or one heavy weapon, and each hex 100 metres. Losses are taken in whole units *and* separate casualties. Tanks, artillery, air-strikes, flak and glider landings included. Map rather bland, since some scenarios involve disregarding terrain. Good tactical stuff – includes confusion! (cv)

AIR FORCE, Battleline (not in polls) Tactical Second World War air game, painstakingly detailed with every aspect of air combat, from permissible manœuvres at different speeds to the chance of baling out if doom strikes. Simultaneous movement, but solitaire scenarios available, though these are necessarily less fun. Very complex and realistic. Length varies with the scenario, from half an hour upwards. (*See* Chapter 8.)

AIR WAR '78, SPI (not in polls) Tactical air game featuring combat with the latest designs, appearing in early 1977.

ALEXANDER THE GREAT, AH (SPI 5·7, 143/AH 6·07, 15) Tactical battle from Alexander's campaign against Darius. Notable for the exotic units (phalanxes, chariots, elephants, archers) and the violent green mapboard, which some find exciting and others off-putting, but nobody has yet called nondescript. Lively positional manœuvres as each side attempts to take the enemy in the rear, thereby slashing defence strengths. Lack of varied terrain concentrates attention on tactical duels and morale levels. Quite easy to learn. 2–3 hours.

ALIEN SPACE, Lou Zocchi (SPI 6·3, 93) Small Star Trek-based game, with eight very different star ships, each with surprise secret weapons unknown to the other players. Game length 15 minutes–3 hours, depending on the number of players.

AMERICAN CIVIL WAR, SPI (SPI 6·1, 109) Grand strategic level, with sea and rail rules and map covering most of the nation. Big-game hunters should look instead at *War Between the States*, or AH's scheduled mid-1977 giant with 2000 counters and eight maps.

AMERICAN REVOLUTION, SPI (SPI 5·9, 128) Strategic level, using large zones rather than hexes. Main emphasis on land war, but British and French navy feature as well as five types of infantry. A dozen optional variations with 'what-if' factors. 'Idiocy rules' hamper the British commander to avoid anti-historical brilliance on his part unbalancing the game. Fairly easy to learn. 2–4 hours.

ANCIENT CONQUEST, Excalibre (SPI 6·5, 65)

ANTIETAM, SPI (SPI 7·1, 7) Part of the successful *Blue and Grey* (I) Quad, and the most popular SPI land game in the poll bar *War in the East*. Special rules limit the number of Union units able to move (bad organization). A close game. (cv-np)

ANVIL-DRAGOON, Jagdpanther (not in polls) Covers the invasion of south of France from beachhead to Grenoble, at regimental level. Units include fleets, airpower, paratroops, commandos, garrisons, and coastal defence forces. Scenarios examine possible D-Day landings, or Anzio in this area. Supply limits the Allied advances. Not a popular game as the German player spends most of the time retreating. Even when he wins he has still lost France! (cv)

ANZIO, AH (SPI 6·4 80/AH 6·64, 2) Not just the battle for Anzio but the war in the whole of Italy, from soon after the first Allied landings until the end. Tremendously detailed long mapboard, with scores of units of different types and strengths on each side. Units are attritioned step by step, but low piece density prevents trench warfare, and sudden breakthroughs are common, especially early on. Four complexity levels from something a bit harder than the classics to ultra-realism. Interesting, varied and challenging, but not for the impatien. 3–9 hours, depending on players' choice of scenario. (*See* Chapter 2.)

ARAB-ISRAELI WAR, AH (not in polls) Scheduled for release as this book goes to press: a development of *Panzerleader*, with air rules further developed and all the modern weaponry available in the Middle East.

ARDENNES OFFENSIVE, SPI (SPI 6·6, 54)

ARMS RACE, DCC–AWA (not in polls) I don't know this company's games myself, though I understand they are attractively produced. *Arms Race* has the US and USSR (and China in a three-player version) slugging it out from 1950 to 2001 with armour, infantry, fighters, bombers, spy satellites, U2s, guerillas, transports, subversive political groups, secret foreign services, extensive production rules, and the threat of nuclear war.

ARNHEM, SPI (not in polls) Part of the most popular Quad, West Wall. A multi-faceted game with paratroops playing the key role, with the Allies trying to link up and the Germans harrying their flanks. (cv-np)

ASSASSINATE HITLER, SPI (not in polls) Unusual multi-player game (with solitaire and two-player scenarios) featuring the power struggles behind the facade of the Third Reich. Released as this book goes to press.

ATLANTA, Guidon (SPI 5·4, 169)

AUSTERLITZ, SPI (SPI 6·6, 54) Simple, well-balanced operational level game.

AVALANCHE, GDW (not in polls) Fascinating struggle for beaches by Salerno with both armies represented at company and sometimes platoon level. Initial beaches are divided by rivers which must be bridged

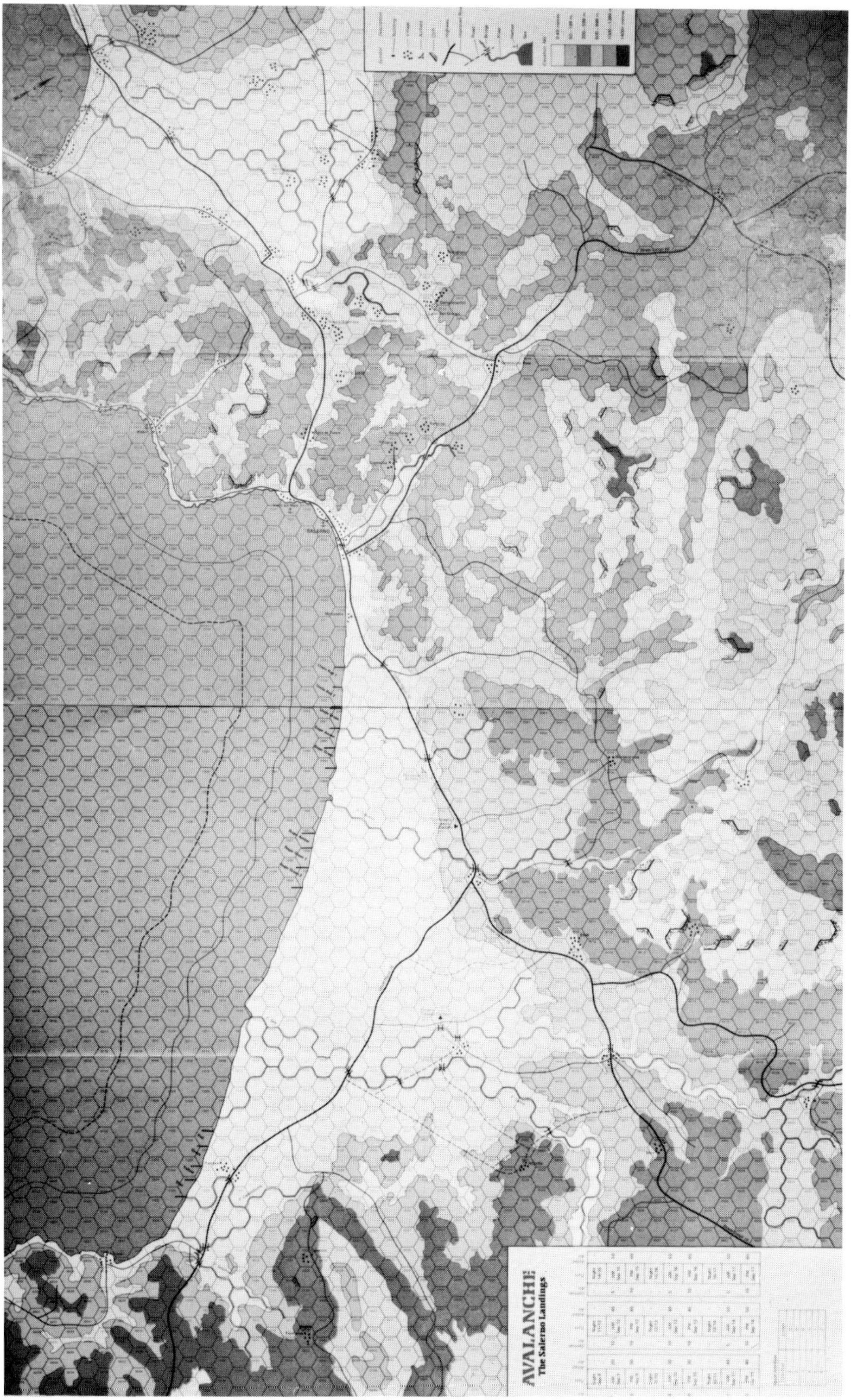

19 Mapboard of *Avalanche.*

by engineers; the bridges can be destroyed by German artillery, but the artillery is silenced by Allied naval gunfire if it groups together. Numerous special-function units, attractive board, clear rules, interesting choice of strategies for both sides, but very long (up to fifty hours or so), and with stacking rules requiring irritating checks on the battalion affiliations of each stack. *Marvellous* for lovers of complex, realistic games, but frightful for a beginner. *See* Chapter 8.

BARBAROSSA, SPI (SPI 6·0, 119) First produced in 1969, but one of SPI's most enduring successes. Fast-moving but fairly realistic; weak ZOCs and a second movement phase enable encirclements and breakthroughs. Four scenarios feature a year each from 1941 onwards, and a fifth indulges the fans of long campaigns by linking the others together. 4–6 hours per game year, intermediate complexity.

BAR LEV, Conflict (SPI 7·1, 7) Seventeen-turn simulation of 1973 Yom Kippur War. Two fronts connected by transit boxes, Golan front at half the Suez scale. Regimental battalion level. Sudden-death unit combat results, with artillery neutralizing or destroying their targets. Complex sequence of play with many different unit types. Heliborne and paratroop raids, SAMs, ECM raids, ground *v.* air attacks, and air superiority feature. Great fun. (cv)

BASIC AIR COMBAT, Lou Zocchi (not in polls) Fighter *v.* fighter ultra-tactical game, with a clever system of energy use to regulate manœuvrability. Eight fighters are included, with eight bombers, against which solo tactics can be practised. Scrambling, evasion through clouds, and fires from non-self-sealing fuel tanks are included, and an advanced version extending the game is on the way.

BASTOGNE, SPI (not in polls) Part of the West Wall Quad, with both siege and relief. (cv)

BATAILLE DE LA MOSKOWA, Martial Enterprises (SPI 7·4, 2) Voted best amateur game in the poll, with an extraordinary second place, this is a simulation of Napoleon's battle at the gates of Moscow which is quite out of the ordinary. Over 1000 pieces, large maps, and highly detailed rules with an emphasis on tactical accuracy. Thus, regiments are more effective in close combat than battalions, but have to break down into the smaller formations to get through hills, and re-forming can be hindered by casualties. Step reduction, powerful morale effects. Very long indeed (40 hours plus) but with satisfying shorter scenarios of 6–8 hours. Very complex.

BATTLEFLEET MARS, SPI (not in polls) Subtitled 'Uprising of the Martian Colonies, 2096 AD', with political, economic, and tactical and strategic military considerations.

BATTLE FOR GERMANY, SPI (SPI 6·0, 119) An imaginative idea, this: the race between the West and the USSR to occupy the largest part of Germany in the closing stages of the war, with each player taking the Germans on the front facing his opponent, or a third (or even fourth) player taking the German side(s). Very fast (1–2 hours), well-balanced between the two main players, and with a simple basic system. Not a bad game for beginners and useful for anyone wanting a quick game, but weak on realism.

BATTLE FOR MADRID, Jagdpanther. Twelve-turn game using a fairly simple move-combat system resembling the Quads. Units include International Brigades, militia, regular Republican forces, some armour, Italian troops, Moroccan tabors, Nationalist forces, and others. The Republicans must hold two fronts to keep the capital. Simple, easy to play, with plenty of room for action. (cv)

BATTLE FOR MIDWAY, GDW (not in polls) Another sizable GDW production, filled with detail and 6–10 hours to play. Unfortunately, many of the interesting features – submarines, initiative, invasion, combat air patrol, air combat in waves, storms – are ineffective or have an odd effect on the game, and the full-length game adds almost nothing to the abbreviated one. Plenty of potential, but purchasers should experiment with rule modifications.

BATTLE OF BRITAIN, Lou Zocchi (SPI 5·6, 150) Three complexity levels: one can be played by children, and takes under an hour. The basic game at adult level includes bases, refuelling, bomber accuracy, flak, diversionary attacks, altitude, radar and navigational errors, and takes 6–12 hours. The advanced game includes weather, aircraft production, strafing, veteran pilots, and generally greater realism. A revised set of rules is available separately; this clarifies ambiguities, and includes yet more realistic rules for the fanatic, as well as a kit for postal play. *See Their Finest Hour* for one alternative.

THE BATTLE OF FIVE ARMIES, Fact and Fantasy (SPI 5·7, 144) Based on the great battle of *The Hobbit*, Tolkien's forerunner to the famous *Lord of the Rings*. Production and rules below usual wargame standards, with few strategic alternatives and the option for one side to

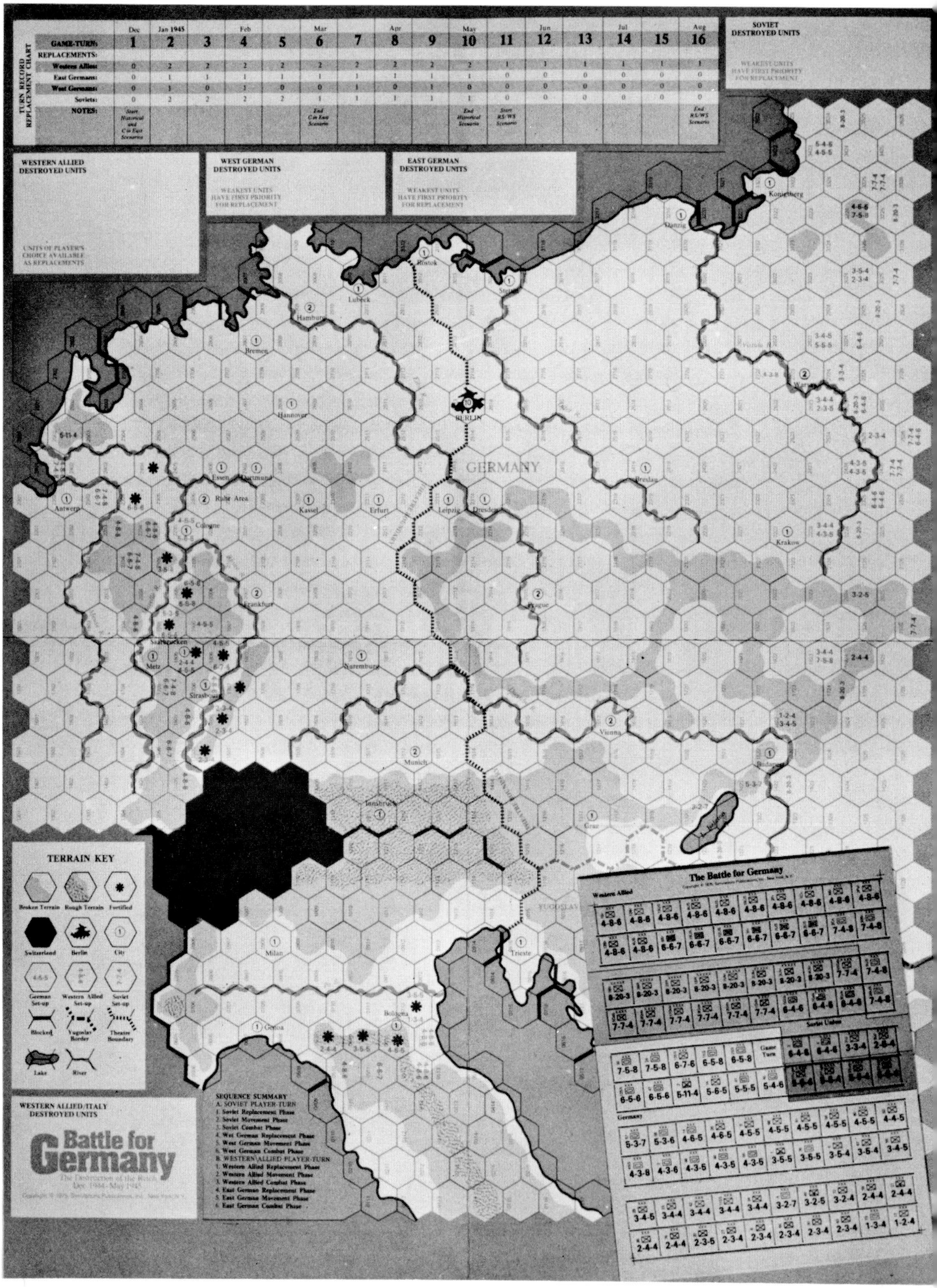

20 Mapboard and units of *Battle for Germany*.

retreat inside a mountain dominating the battlefield, and engage in a lengthy die-rolling contest as the enemy tries to force his way inside. Not recommended except to keen Tolkien fans, and even they are advised to modify the rules. Simple to learn and play. 2–5 hours.

BATTLE OF NATIONS, SPI (SPI 6·9, 23) Part of the *Napoleon at War* Quad. The wrong scale for the battle, but a very exciting encirclement struggle. (cv)

BATTLE OF THE ATLANTIC, DCC–AWA (not in polls) See *Arms Race* re DCC–AWA. This game is strategic, with varying numbers of wolf packs, Condors and Luftwaffe raiders facing battleships, cruisers, carriers, escort carriers, frigates, destroyers and convoys. Sixty-plus counters, two maps, physical quality said to be less good than usual in wargames.

BATTLE OF THE BULGE, AH (SPI 5·9, 128/AH 5·79, 20) I have a problem here: this game is one of my favourites, but the polls suggest this to be very much a minority view! Regimental level, with enormous German impetus gradually stemmed by US reinforcements. The special charm of the game is that it is usually almost impossible to tell who is winning for a long time, as the German progress always looks spectacular – but the US will turn the tide if the advance is not sufficiently successful. In my view, the game is almost balanced, given vigorous defence in depth by the US, but most games show a big German advantage. This, plus problems with certain rule interpretations and gaps in realism, add up to the low poll rating, but it remains frequently played and guarantees an exciting time. Intermediate difficulty, 4–6 hours. (*See* Chapter 6.)

BATTLE OF THE MARNE, SPI (SPI 5·5, 159) Two-scenario game with each side's August 1914 offensive. Quick set-up, easy to play, rather simple map, and not a very large number of units, so one of the shorter games (2–4 hours). The lack of complexity makes it perhaps a little *too* straightforward to attract great interest.

BATTLE OF THE WILDERNESS, SPI (SPI 6·4, 80) The least popular of the successful *Blue and Grey* (II) Quad. The battle is limited to the roads and tracks in the wooded areas. (cv)

LA BELLE ALLIANCE, SPI (not in polls) Pitched battle at the climax of Waterloo; part of the *Napoleon's Last Battles* Quad.

BAY OF PIGS, Jim Bumpas (not in polls). Operation simulation of the 1961 invasion of Cuba, by the designer of the popular *Schutztruppe*. Amphibious landings, unit disruptions, supply limitations, air and naval bombardment. Fairly easy to play and fast-moving. The game was dedicated to the fifteenth anniversary of the successful Cuban defence.

BLITZKRIEG, AH (SPI 5·7, 144/AH 5·91, 17) Ambitious attempt to incorporate every aspect of modern warfare in an abstract context does not quite come off; both sides have very similar mixes of infantry, armour, artillery, paratroops, rangers, fighters and different bomber types, and most players steer the game into boring wars of attrition. Given aggressive play on both sides, however, the game comes alive with a bang, as there are liberal opportunities for automatic victories allowing constant breakthroughs and encirclements. Not recommended by post (too many units) or solitaire (player interaction is needed to make things hot up). 5–8 hours, intermediate complexity.

BLITZKRIEG (MODULE SYSTEM), SPI (SPI 6·4, 80) Unique example of one leading company building on the game of another. The module system is only usable in conjunction with AHs *Blitzkrieg*, and adds a number of new CRTs for different types of air and land warfare, plus a naval combat one. 700 new counters, including minor country forces, and rules for railways, three types of movement, production, weather, guerillas and other matters. Any attrition tendency in the original game is removed, and the result is widely felt to be an improvement, though naturally more complex. *Strategy I* owners will recognize some of the techniques. Game length still 5–8 hours.

BLOODY RIDGE, SPI (SPI 6·5, 65) Part of the *Island War* Quad, with a fast-moving and clean-cut game on the battle for Guadalcanal, concentrating on the area around Henderson Field. Well balanced; the Japanese have the onus of the offensive, but a tactical edge with an ability to infiltrate and disengage. (mw)

BLUE AND GREY (I), SPI (SPI 6·9, 23) The second most popular Quad, just behind West Wall. See *Antietam*, *Chickamauga*, *Shiloh*, and *Cemetery Hill.* Simple tactical surround-and-destroy system. (cv)

BLUE AND GREY(II), SPI (SPI 6·6, 54) Not quite such a hit but still favoured. See *Hooker and Lee*, *Chattanooga*, *Battle of the Wilderness* and *Fredericksburg*.

BORODINO, SPI (SPI 6·9, 23) Operational-level simulation of Napoleon's qualified victory which tempted him to advance to Moscow. Very simple and fast-moving game, based on the introductory *Napoleon at Waterloo*, yet durably popular with the hard-core. Only 91 units; 1–3 hours. For the same battle with an ultra-complex treatment, *see La Bataille de la Moskowa.*

BREAKOUT AND PURSUIT, SPI (SPI 6·3, 93) Title suggests some revoltingly abstract game, but in fact this deals with the battle for France after the D-Day landings (not themselves featured). A sixteen-turn campaign game covers the full July to September period, and scenarios feature individual parts, from the 'breakout' from Normandy to the 'pursuit' race for the Rhine. Emphasis on Allied logistical problems. Fairly complex, 3–5 hours.

BREITENFELD, SPI (not in polls) Good brisk game using the system of the *Thirty Years War* Quad.

BULL RUN, SPI (SPI 6·1, 109) Operational level, with hidden *and* simultaneous movement, hence not for the weak-minded. Five scenarios.

BUNDESWEHR, SPI (not in polls) Nato *v.* Soviet bloc forces. Part of the *Modern Battles II* Quadrigame.

BURMA, GDW (not in polls) Not as many units as usual from GDW (240), but still not for the coffee-break, with twenty-six turns and quite complex rules; 5–10 hours once one knows them. Mostly brigade/regiment level, with a turn per month from December 1942. Suitably jungley-looking map with lots of difficult terrain, which helps the thin Allied defences to stop a Japanese breakthrough. Complex supply rules, and the usual intriguing GDW special rules: glider-borne Long Range Penetration Forces requiring special training (can they be spared from the front?), engineers working away at the Burma Road, and Chinese reinforcements of questionable enthusiasm. Fairly bloodless CRT; a game of manœuvre.

CA, SPI (SPI 6·3, 93) Tactical Pacific warfare at sea around Guadalcanal, with battleships, cruisers and destroyers. Ten scenarios. Popular with naval fans, but an older design than the relatively recent successes in this theatre: *Dreadnought*, *Frigate* and *Wooden Ships and Iron Men.*

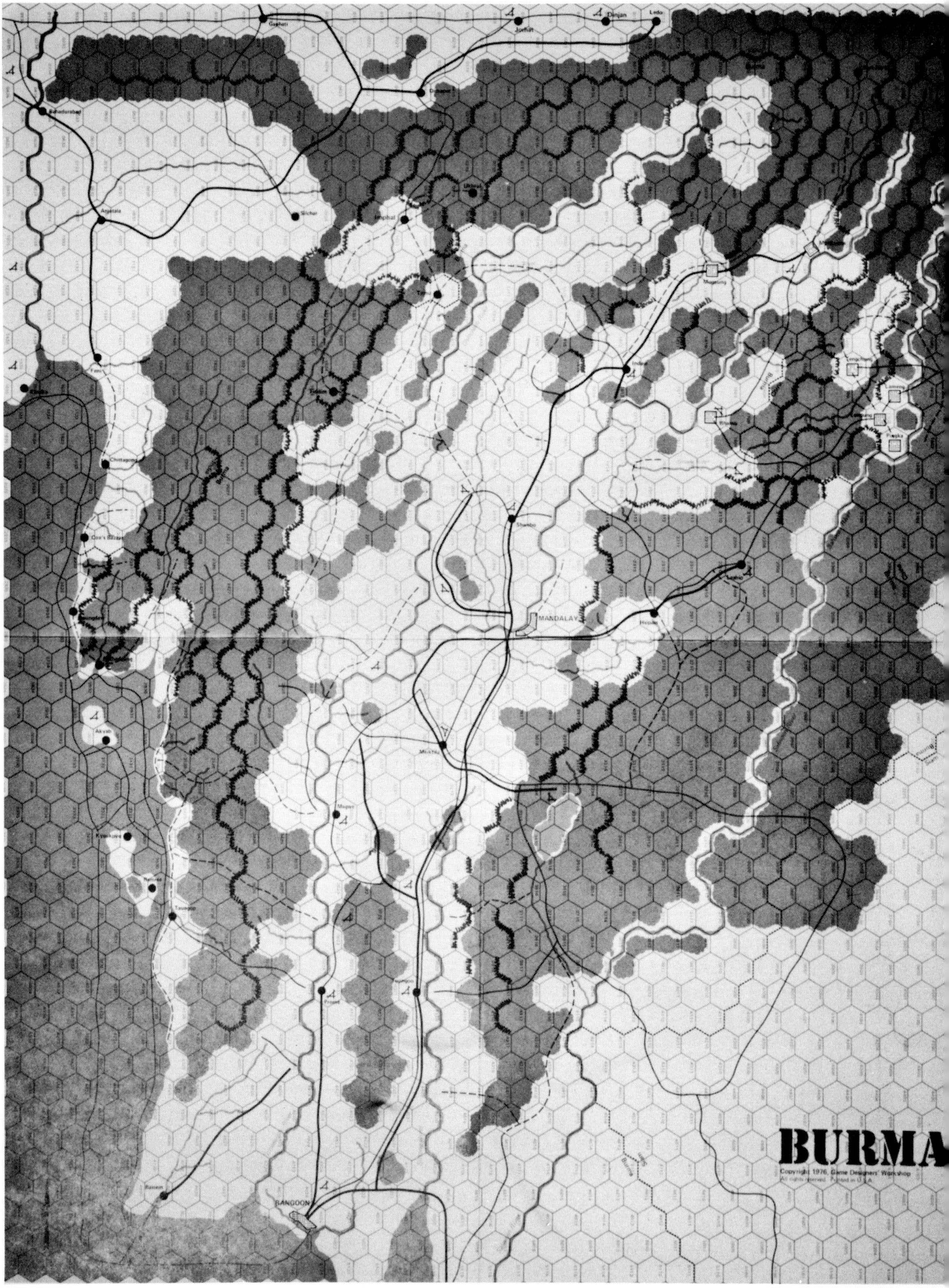

21 Mapboard of *Burma*.

CAESAR, AH (not in polls) Much-admired, previously amateur, game on the battle of Alesia, a remarkable double siege during the Gallic Wars. The Gauls under Vercingetorix are holed up in the fortified city of Alesia; Caesar is ensconced in twenty-five miles of fortifications, constructed by him to face both ways, around Alesia; quarter of a million Gallic reinforcements have arrived outside the fortifications. Ten legions, Germanic cavalry, Balearic slingers, Numidian archers and light infantry, numerous forts, and inspired leadership give the Romans a fair chance despite a 6–1 numerical inferiority. Playable by 2, 3 or 4 participants in 4–5 hours; moderately complex, with a rich variety of units.

CAESAR'S LEGIONS, AH (SPI 6·0, 119/AH 6·36, 9) Low rating conceals a wide range of views on this tactical game between Romans and German tribes. Scenarios run from the simple to the rather complex, a technique used to an even greater extent in *Tobruk*. The advantage is that newcomers can work their way gradually into the rules, instead of learning page after page before moving a piece. Unfortunately, the early scenarios are too simple for most tastes, and unbalanced in favour of the Romans. The later ones, however, are absorbing and varied. German play is always tricky, involving hit-and-run guerilla tactics. Combat is modified by 'tactical cards', also seen in *Kriegspiel* and *1776*; each side chooses between various stances ('refuse the left', or 'withdraw' are two examples), and the interaction modifies casualties. This feature makes the game not altogether suited to solo play, but it should appeal to anyone interested in the period, even if the Latin names are a trifle distorted at times!

CAMBRAI, Rand (SPI 5·5, 159) Simple and exciting games are what Rand tries to produce, and *Cambrai* seems to meet the target, as well as being well-balanced. Each side has a period of 'surprise' in which they can use a special CRT to gain the upper hand. The low poll rating of most Rand games is probably partly due to their relative simplicity, and partly to some rule obscurities, so it is wise to ensure that you have any errata sheets around. The Rand games can be bought separately or in a large and (per game) cheap package. The company has had financial difficulties in 1976 and the games are now hard to obtain.

CAULDRON, SPI (not in polls) The Battle of Gazala, May 1942. Part of the North Africa Quad. Much longer (twenty-six turns) than usual Quad games, with control of Tobruk crucial.

CEMETERY HILL, SPI (SPI 6·8, 33) Slightly less popular than the others in the first *Blue and Grey* Quad; a rather bland Gettysburg. (cv-np)

CHACO, GDW (SPI 6·4, 80) In total contrast to every other GDW game, this has a scenario only two moves long! The main scenarios are also short and playable, the theme being the little-known Bolivian–Paraguayan war in the 1930s, which led to an appalling bloodbath as First World War techniques were married to the Second World War firepower. The basic game is simple and brisk; the advanced one consists of a choice of eleven optional rules; each player chooses the rules he would like to use, plus the rule he *bars* the other player from choosing, an interesting touch. A peculiar variant introduces the US Marines.

CHANCELLORSVILLE, AH (SPI 6·2, 102/AH 6·32, 9) An early AH production now much improved in a new edition. Attractive board; detailed rules for river crossing; bloodless CRT, with units dispersed and regrouping, often during the night; optional artillery rule and other possibilities. The game is generally thought well-balanced, with an edge to the Union; as with *Bulge*, the difficulty in seeing who is winning is that the Union always does splendidly, but needs a massive triumph to win the game. 3–6 hours, quite complex.

CHARIOT, SPI (SPI 6·0, 119) Part of a series of eight games depicting combat from 'the dawn of civilization' up to 1900, of which the five up to 1550 form a group which I suppose should be called a Quintigame. This group, known as the Prestags (Pre-Seventeenth Century Tactical Game System), have the same standard rules, with relatively minor modifications for special features in each period. *Chariot* is the first, covering the Biblical period, followed chronologically by *Spartan* (Greek era), *Legion* (Roman era), *Viking* (Dark Ages, including the Crusades), *Yeoman* (Renaissance – Bannockburn, Agincourt, etc.). The Prestags can be obtained together in a 'Master-Pack', and are compatible with each other, so you could even have chariots charging longbowmen if you wish, and can persuade some sap to play the chariots. The later tactical games, *Musket and Pike* (1550–1680), *Grenadier* (1680–1850) and *Rifle and Sabre* (1850–1900), use a different though similar approach. *Chariot* has not been as well received as some in the poll, probably because it is one of the simpler ones, and there is little surviving literature to suggest special features about warfare in the period.

All the games in the series emphasize tactics, and the maps are mostly nondescript, with great expanses of clear terrain and a few hills, rivers, and other features dotted about. The variation comes in the different unit types, and the numerous scenarios in each games. Game length is rather shorter than usual (2–4 hours) for most of them. The function of the series, tactical clashes with a simple basic system, is well achieved, though the anonymous maps inhibit the usual war-games enjoyment of refighting a particular battle. As an aside, it may be added that *Chariot* is a revised version of a game enticingly called *Armageddon*, but neither game actually attempts to simulate the Biblical Armageddon itself, despite considerable theological controversy on the stacking limit for angels on a hex-per-pinhead scale. Pity.

CHATTANOOGA, SPI (SPI 6·6, 54) Part of *Blue and Grey* (II). Union troops try to battle out of the siege. (cv)

CHICKAMAUGA, SPI (SPI 7·0, 15) Part of *Blue and Grey* (I). Fought in forest with a great deal of manœuvre and road-blocking; lots of options. (cv)

CHINESE FARM, SPI (SPI 7·0, 15) The most popular of the successful *Modern Battles* Quad in the poll, this has nothing to do with Chinese farms and deals with the Arab assault on Israel in 1973, which also features as one scenario of SPIs *Sinai*, which has an equal placing with *Chinese Farm* at the top of the lists, but covers all fronts. A drawback about *Chinese Farm* is that the CRT is unsuited to cross-canal actions. (cv-np)

COMBINED ARMS, SPI (SPI 4·4, 195) Disastrously unpopular game featuring tactical operations in the 1939–1980 time span. It seems better to choose one of the numerous other games in this general area: *MechWar '77*, *Panzer '44*, *Panzerblitz*, *Panzerleader* and *Kampfpanzer* spring to mind; *Tobruk* is also good for combined arms operations.

THE CONQUERORS, SPI (not in polls) Two basically strategic simulations of Alexander's Persian campaigns and the Roman drives in the Mediterranean, published in early 1977.

CONQUISTADOR, SPI (not in polls) A complex and highly entertaining game for one to three players based on the exploration and conquest of the New World in the sixteenth century. England, France and Spain wrestle with each other and with all manner of amazing events, from soldiers demanding to look for El Dorado to Caribbean Indians going

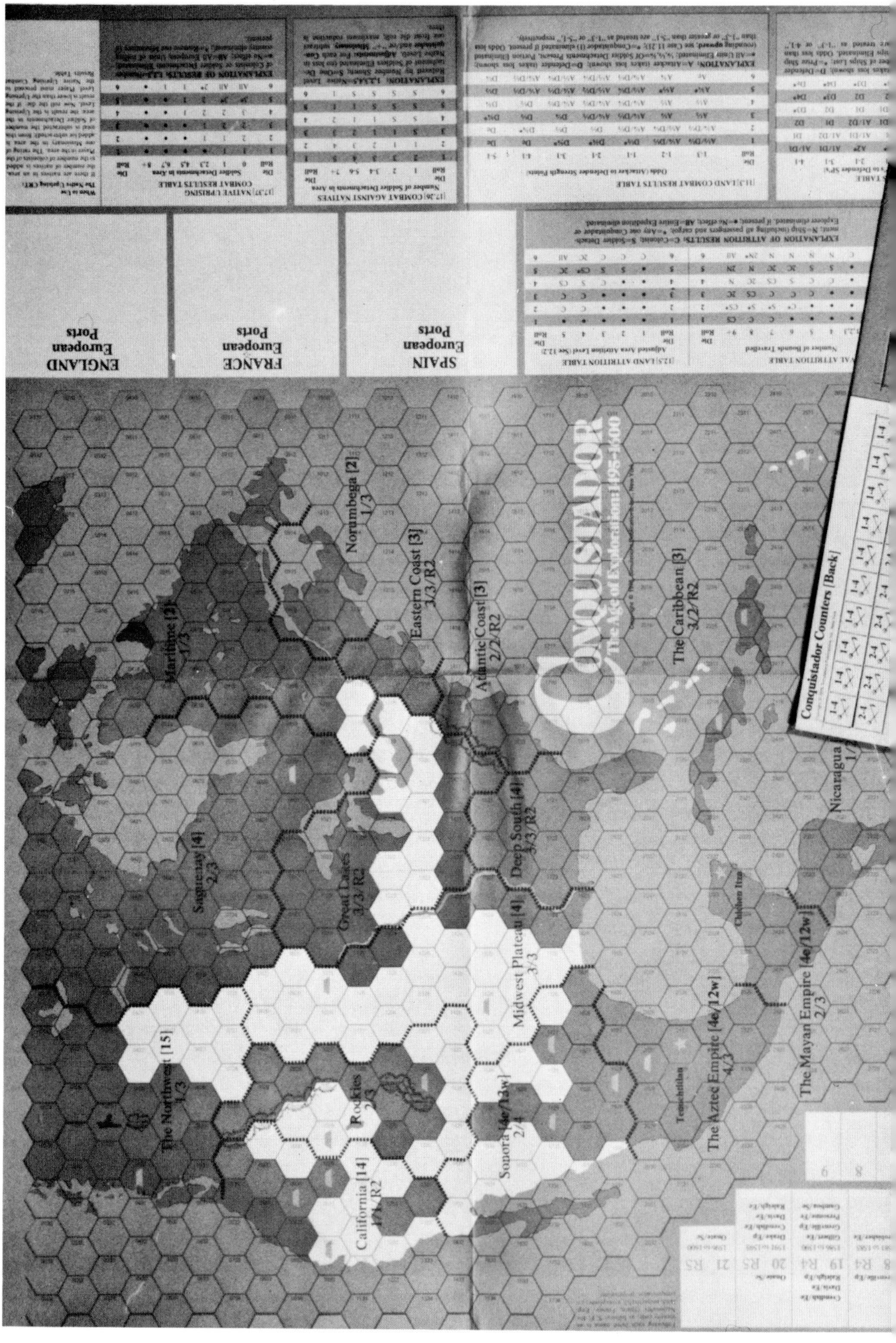
CONQUISTADOR
The Age of Exploration 1495-1600
Conquistador Counters [Back]
ENGLAND
European Ports
FRANCE
European Ports
SPAIN
European Ports
The Northwest [15]
1/3
Saguenay [4]
2/3
Maritime [2]
1/3
Norumbega [2]
1/3
Eastern Coast [3]
3/3/R2
Atlantic Coast [3]
2/2/R2
The Caribbean [3]
3/2/R2
Great Lakes
3/3/R2
Deep South [4]
3/3/R2
Midwest Plateau [4]
3/3
Rockies
3/3
California [14]
1/1/R2
Sonora [4e/12w]
2/4
The Aztec Empire [4e/12w]
4/3
The Mayan Empire [4e/12w]
2/3
Tenochtitlan
Nicaragua

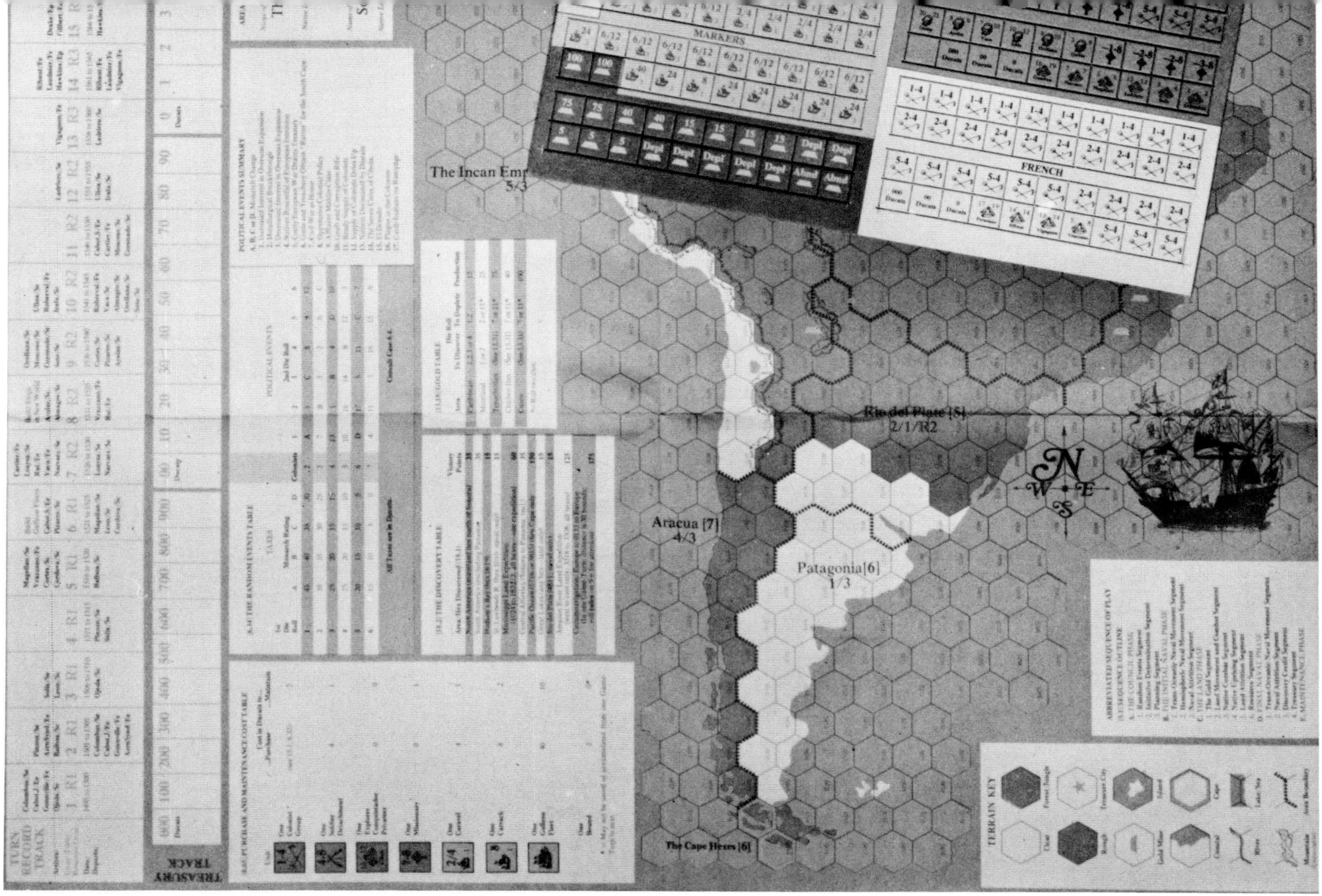

22 Mapboard and units of *Conquistador*.

on the rampage. Expeditions, diplomacy, South Cape navigation problems (requiring a 'Rutter' guide), missionaries, privateers, colonists, natives and (naturally) conquistadors feature in what is very much a 'fun' game reminiscent of but even more varied than *Kingmaker*. Not a supreme test of skill.

CORAL SEA, GDW (SPI 6·5, 65) Companion game to *Battle for Midway*, depicting the first carrier battle in the Pacific; one can base a *Battle for Midway* game on the results of a *Coral Sea* game, though fans of Pacific campaigns may prefer the massive *Pearl Harbor* which John Prados of *Third Reich* fame is bringing out in 1977 after this book goes to press, or SPI's *War in the Pacific* or AH's equivalent, both of which are due in mid-1977. *Coral Sea* is not one of GDW's megagames, and has just 240 counters and a 22″ by 28″ map. Unusually, naval movement is hidden, but air movement is open. Highly detailed ship and air descriptions, aircraft maintenance, coast-watchers, and (to wake up sleepy coast-watchers) invasions are featured. The game reviewed here is the second edition, which has made some minor additions and modifications. Air counters follow the *Europa* pattern, and 90% of the ship and air counters are interchangeable with *Battle for Midway*. Noted for realism, but rules require concentration. 6–10 hours.

CRIMEA, GDW (SPI 7·1, 7) A very rarely simulated theme, despite the famous Charge of the Light Brigade, but this game's enthusiastic reception from the hard-core demonstrates the potential interest of the subject. 480 attractively printed counters in five colours; map unusually chopped into separate sections, well-printed but a little less detailed than usual from GDW. The basic game concentrates on West Crimea (the largest map section), while the advanced version brings in the other areas. Innovative combat system reinforces correct period 'feel'. Another delight for aficionados, intimidating for beginners.

CROMWELL, SDC (not in polls) Quietly distinguished simulation of the English Civil War, with a simple basic system which is gradually elaborated in more advanced versions. Strategic level, with big zones; large armies roam the countryside trying to widen their spheres of influence for the winter recruiting phases. Strong leader effects; fleet rules; Scottish intervention. Not nail-bitingly gripping, but consistently interesting and challenging. *See English Civil War* for an alternative.

CRUSADER, SPI (not in polls) Tobruk, November 1941. Part of the North Africa Quad. Emphasis on mobile warfare, as the British try to relieve the fortress. Two short scenarios and a campaign version.

CUSTER'S LAST STAND, Battleline (not in polls) Battleline have made their name with fine tactical games accenting simulation and accuracy, but *Custer's Last Stand* is a venture into a new area for wargames with a rather more strategic flavour. Custer approaches a large Indian encampment which is sitting peacefully behind a river sending out small patrols. When they spot each other, Custer sends off a 'runner' to get reinforcements, and the camps start to stir uneasily. Custer can choose to hang about waiting for powerful reinforcements, or he can make an immediate foray, with the chance of some victory points but a very grave danger of repeating history. If he chooses prudence, the game develops into an absorbing pursuit of the Indians as their war parties try to screen the withdrawal of the camps (who presumably include their families, so one's sympathies tend to be with the Indians). Interesting choice of strategies, but some rule problems. 4–6 hours, moderately complex.

D-DAY, AH (SPI 5·4, 169/AH 5·57, 21) Not terribly popular in either poll, but this AH 'classic' has quite a few devotees, and it often appeals to beginners. The accent is on a fast, easy game rather than on historical accuracy – with all Allied units except paratroops much of a muchness the game is indeed not very accurate.

However, as in most of the 'classics', the historical 'feel' and the choice of strategies are conveyed rather well. The Allies have a choice of invasion areas, and will generally (if they're sensible) choose one which is lightly-defended, even though this is likely to be a long way from the goal of the Rhine. A counter-attack on the beaches will probably fail, so the Germans are pursued across France until they turn to fight in the last fortification line in front of the Rhine, where they have a chance to hold out. Widely thought to be vastly biased to the Allies, but a conservative German defence can work well; alternatively players can choose to have the initial German units turned face down, my personal variant for enlivening the early turns. 3–6 hours.

DECLINE AND FALL, Wargames Research Group (SPI 5·8, 137) One of the very few British-produced games, this is disliked by experts because of its elementary rules and unbalanced structure, but it is a highly entertaining game for four players, if it's not taken too seriously. A powerful Rome sees its static frontier forces under assault from the

Vandals and Goths, and has too few mobile forces to altogether cope, particularly when the Empire splits into East and West during the game. The more territory the German tribes occupy, the faster they recruit more fighters to the banner, so each tries to occupy safe 'breeding grounds' in remote parts of the board. The wild card is the Hun player, whose forces grow on blood-lust, with more units coming on when reports of enemies destroyed flow east. The weak Germans need Hun help to dish the Romans, but to prevent the Hun then eating them up they usually have to call hopefully on the Gods to strike the Khan dead! May be produced by a US company before long. 3–5 hours. Note that although the Wargames Research Group are normally a high-quality miniatures group, with the popular miniatures-based *Seastrike* to their credit, *Decline and Fall* is not at all reminiscent of miniatures. Both games are now marketed by Philmar.

DESERT WAR, SPI (SPI 6·5, 65) Companion game to *Kampfpanzer*, featuring platoon-level North African combat with simultaneous movement.

DESTRUCTION OF ARMY GROUP CENTER, SPI (SPI 5·5, 159) Massive Soviet offensive unleashed on still powerful German defences; in real life, half a million German troops were killed or captured as an enormous pincer sliced through their flanks. Something of a slogging match without much chance to change strategies after the initial deployments, which should be made with corresponding care. Simple, smooth-running game system makes 'DAGC', as it is usually called, a good game for beginners, but experienced players tend to find it lacks variety. 3–5 hours.

DIEN BIEN PHU, SDC (SPI 4·2, 200) This gruesomely-rated game is rarely seen now, though the subject must have looked promising enough. It is accurate and innovative but unbalanced against the French (cv-np).

DIPLOMACY, AH/Philmar (SPI 6·8, 33) Although Avalon Hill now distributes *Diplomacy*, it is not a wargame in the usual sense of the word, as it has zones rather than hexes and extremely primitive movement and combat rules. However, the game has an extraordinary charm all its own, based on the fascinating interplay of alliances between the seven players representing different European powers at the turn of the century. The rule-book specifically encourages double-dealing and back-stabbing, and postal players in particular refine this to an art which Machiavelli would have appreciated, and dozens of amateur magazines run games, with over 200 national and international games started by the National Games Club alone and over a thousand in the US hobby. The search for a true wargame with the same diplomatic interest continues (*Strategy I*, *Origins of World War II* and *Third Reich* do not quite manage it); in the meantime, this game can be strongly recommended to anyone interested in games of multi-player negotiation. 4–10 hours, or 18 months–2 years by post, with moves made once a month. (*See* Chapter 5.)

DIXIE, SPI (SPI 4·9, 184) 68% of the SPI sample knew this game and on the whole gave it a negative rating. The idea itself has a zany appeal: the Confederacy is assumed to have *won* the Civil War, and the two sides, now separate countries, are squaring up for a rematch in the present age with modern technology. This gives an excuse for having a game fought with modern weapons on familiar American soil, but *Invasion: America* was better received.

DMZ, SPI (not in polls) Part of the *Modern Battles II* Quad, set in a hypothetical contemporary clash in Korea.

DRANG NACH OSTEN/UNENTSCHIEDEN, GDW (SPI 7·5, 1) The winner of the SPI poll and the first installment of the hard-core's dream: GDW's Europa series. Statistics are the best way to describe the games. *DNO* has five 21″ × 27″ maps, 1700 counters, fifteen pages of rules and charts, and covers the Soviet front of the Second World War from 1941–2 at a divisional level. *Unentschieden* is (I kid you not) for those who find *DNO*s scope insufficiently broad for their tastes, and includes four further half-maps, so that the total map runs from the Polish border to Archangelsk and Kubishev, as well as a further 1900 counters (some of them admittedly replacements) and fifteen pages of rules and charts covering the war to 1944–5. However, GDW would hate to cramp anyone's style, so the full Europa series will consist of a mind-boggling twenty games (plus extra modules), with a total map size covering thirteen feet square, though the maps will overlap slightly, and allow gaps between theatres of war. The final masterpiece will cover the entire war in Europe, and would probably take as long to play as the real thing if only two players were involved. *Narvik* and *Their Finest Hour* (the Battle of Britain) are the other games in the series already out (though *Narvik* does not yet usefully slot into *DNO*, since the Norwegian campagn was earlier), with *Case White* (the Polish occupation, 1939) and *Marita-Merkur* (the Balkans and Crete, 1941) on the way. There is a school of thought which holds that the Europa series is for admiring and revelling in the rules rather than playing, but it *is* playable, and there is every reason to believe that the final version will be too, given enough time or (more realistically) enough players. The best chance of carrying out a full version of the total series will probably be a postal game with a score of players taking commands at different levels and in different parts of Europe. If it ever happens, I mean to be one of them.

DREADNOUGHT, SPI (SPI 7·1, 7) The SPI poll seems to show considerable enthusiasm for sea games as such, but there is no doubt that *Dreadnought* is an excellent game, based on tactical battleship (plus escorts) combat from 1906 to 1945. Rather unusually in a naval game, the accent is on playability rather than enormous detail, and the result is fast-moving and frequently tensely exciting. Battleships from all over the world (even Brazil and Argentina) feature. Plenty of historical scenarios, while players interested in increasing the skill content can build their own fleets, and divide them into task forces for different types of conflict. (*See* Chapter 8.) 2–6 hours, longer for campaign scenarios.

EAGLE DAY, Histo Games (SPI 5·7, 144) The Battle of Britain. For better-known alternatives, see *Battle of Britain* and *Their Finest Hour.*

EAST FRONT, Control Box (available from Zocchi) (not in polls) Several hundred hex-shaped counters and two large maps, and nine scenarios, ranging from the equal shortest in wargaming, a two-turn invasion of Poland (the other two-turn game is a *Chaco* scenario) to a 101-turn struggle for the Soviet Union.

THE EAST IS RED, SPI (SPI 5·5, 159) Operation-level simulation of a hypothetical Sino–Soviet war in the present era, with the emphasis on conventional warfare, but provision for nuclear strikes. Various scenarios. Good brisk game with little innovation over the *Kursk* game system but unusual theme. (cv-np)

1812, SPI (SPI 5·7, 144) One of the most ambitious projects from SPI a few years ago, *1812* is really two games on the same theme, Napoleon's war in Russia. One is a fairly conventional treatment at a 'grand tactical' level, while the other is a strategic game with large zones instead of hexes, and an innovative combat system. Each game has its own map, counters and rule book, and each has three scenarios based on different periods. Both games are in the 3–5 hour bracket, though the zonal version is a little shorter than the more detailed hex game; both are moderately complex. The package failed to excite great interest, however, and is not often seen now.

EL ALAMEIN, SPI (SPI 5·1, 179) Unenthusiastically received simulation of three months' fighting in 1942, ending with El Alamein itself. The 250 units are at regiment/battalion level, with a necessarily rather blank map. Special features are mines, engineers, and a detailed supply system.

ENGLISH CIVIL WAR, Ironside (not in polls) Area movement campaign. Units come in infantry/cavalry strength points. Each area is coded to show the number of Royalist, Parliamentarian and neutral troops (who join the controlling party) that the area will raise each winter. Leaders, sieges, and the Scots feature. The game is rather long but quite accurate, with a very good tactical module. See *Cromwell* for a not dissimilar alternative. (cv-np)

FALL OF BATAAN, Jagdpanther (SPI 4·7, 189) Dubious reception by a small sample.

FALL OF TOBRUK, Conflict (SPI 6·4, 80) Pleasant tactical North African game by the makers of *Bar-Lev*. Fluid movement/combat system; minefields and engineers feature.

FAST CARRIERS, SPI (SPI 6·4, 80) Very big game with four Pacific scenarios, Korea, Vietnam, and the Denmark Straits. Very accurate if rather slow. Searches, re-fuelling of aircraft, deck space, misinformation, radar, submarines all appear. Many counters, complex system, *but* cleanly designed. Only for the real naval buff. (cv)

FIREFIGHT, SPI (not in polls) Ultra-modern small unit combat, with counters for individual vehicles and groups of four infantrymen. Unusually detailed rules, with particular care taken to see that beginners can follow them, and a massive accompanying handbook with analyses both of the game and contemporary armament and US/Soviet tactical doctrine in real life. Relatively elementary for a tactical game, with such questions as morale and intelligence omitted and a limited range of vehicles, but most impressive within its scope and particularly suitable for beginners interested in modern tactical warfare but anxious to avoid excessive complication. 1–3 hours per scenario, fought on two large and attractive maps.

FLYING TIGERS, Lou Zocchi (SPI 5.0, 182) Based on the exploits of General Chennault's volunteers against the Japanese from December 1941 to July 1942. Includes a 'what-if' scenario involving the Japanese naval air power.

FORMELHAUT II, DCC–AWA (SPI 5·1, 179) Tactical galactic-era space combat, featuring gravitational wells, nebula, null guns, and a 28″ × 24″ five-colour map. Thirteen scenarios, from two-player two-hour games to an operational-level game including ground combat with room for up to thirty-four players. Second edition now in production contains minor corrections.

FOUR ROADS TO MOSCOW, AH (not in polls) In a Quad game, the rules are similar but the board changes; here the 44″ × 22″ map of Russia is used in four simulations of invasions by Mongols, Charles XII, Napoleon and Hitler, with weekly turns, and each campaign studied for up to six months.

FOXBAT AND PHANTOM, SPI (SPI 6·4, 80) Tactical air duels between fourteen modern fighter types. As with most games of the type, the main interest focusses on exploring the strengths and weaknesses of the different aircraft, and games are fast and simple. A short time limit per move is a good idea, to get a little supersonic 'feel'; to spend (as some players do) thirty minutes pondering each move is more suggestive of a duel with pikes by exceptionally ponderous peasants in a swamp.

FRANCE 1940, AH (SPI 6·5, 65/AH 6·18, 10) The problem in designing a game on the fall of France is that anything resembling the historical collapse will fail to appeal to the French player. *France 1940* copes with this by having alternative orders of battle for each side, to enable players to find the balance which suits them best; the historical balance, called the 'Idiot's Game', is included as one option. The standard rules concentrate on playability, with optional additions for greater realism. Fine physical quality, quite difficult rules, length 3–4 hours.

FRANCO-PRUSSIAN WAR, SPI (SPI 5·5, 159) Operational level simulation of the war which alerted the world to Prussian strength. Elementary form of hidden movement (counters are visible but inverted); units reduced step by step, with different CRTs for each defending strength level. In the advanced rules, the order of battle varies, reinforcing the hidden-movement uncertainty. Fairly complex, 2–3 hours.

FREDERICKSBURG, SPI (SPI 6·5, 65) One of the less popular games in the *Blue and Grey* Quads, though still rated well above average. It can be combined with *Hooker and Lee* to simulate the whole Chancellorsville campaign. (*See also Chancellorsville.*) (cv)

FREDERICK THE GREAT, SPI (SPI 6·0, 119) This features the main campaigns of Frederick during the Seven Years War in a number of scenarios. Defenders hole out in fortresses and get besieged, leaders inspire the troops, and prisoners can be taken. This relatively obscure European war was popularized (if that's the word I want) as part of the hero's life in the Oscar-winning movie *Barry Lyndon*, where it is accurately described as 'difficult to understand'. The conflict did have important spin-offs in North America, culminating in the storming of Quebec by Wolfe, for which see *Quebec 1759*.

FREIBURG, SPI (not in polls) Unsatisfactory component of the otherwise excellent *Thirty Years War* Quad; unless you have an errata sheet the campaign game is unplayable due to certain baffling misprints, and even if assumptions are made to fill the gaps, the game remains a series of slogging matches, though it has the advantage of varied terrain, and can generate a fair amount of excitement.

FRIEDLAND 1807, Imperial (SPI 6·4, 80) Uses a Quad-like system, *but* they include HQ/leader units and an eminently sensible CRT. Morale rules used and artillery handled separately. A great improvement on the Quads and worthy of transplanting. Battle hinges on an early Russian victory or a careful retreat over a few bridges. (cv)

FRIGATE, SPI (SPI 7·1, 7) Ship-to-ship combat in the pre-steam era, with multiple scenarios involving British, Spanish, French and Dutch. See *Wooden Ships and Iron Men* for an equally impressive alternative.

FULDA GAP, SPI (not in polls) Subtitled 'the first battle of the next war', with battalion-level combat including electronic warfare; a mobile Nato defence against a multi-echelon Soviet attack.

GETTYSBURG, AH (SPI 4·7, 189/AH 5·12, 24) One of the oldest wargames still in print, *Gettysburg*'s feeble rating in both polls stems from the great advances in game design which have been made since its first appearance in 1958; moreover, it is too simple to appeal to the hard-core, there is a play-balance problem, and the attractive map is insufficiently used. However, it is quite a reasonable game in its class: a brisk, easy classic, and quite a pleasant introduction to the hobby for anyone interested in the period. Experts will prefer *Terrible Swift Sword* or AH's forthcoming advanced game on the battle.

GLOBAL WAR, SPI (SPI 7·0, 15) One of the giant-sized species beloved of the hard-core, with 1200 counters and a two-sheet map, based on the entire Second World War, and consequently lengthy (twenty hours plus), though there are mini-games. Ingenious production system introduced here for the first time. A tremendous challenge to both sides, and as usual in this sort of game well suited to multi-player groups. It is desirable to have the errata sheet. Defence is thought rather overfavoured by some players, making a German invasion of the Soviet Union or an Allied liberation of France an imposing task against good play. Highly complex.

GOLAN, SPI (SPI 6·8, 33) The Golan front of the 1973 Arab-Israeli war, brother of the still more popular *Chinese Farm* in the *Modern Battles* Quad. Rather unbalanced to the Israelis. (cv-np)

LA GRANDE ARMEE, SPI (SPI 6·6, 54) The most popular operational level Napoleonic game in the poll. Rather bare map, but four hundred varied counters from different nations featuring in well-designed scenarios from 1805 to 1809. The successful formula is a simple basic system giving the special problems of war in this period: depots and supply; cavalry screening; fortresses; forced marches; individual tactical abilities of leaders. Game length 3–4 hours, rules as a whole moderately complex.

GRENADIER, SPI (SPI 6·1, 102) See *Chariot* for a description of the ideas behind SPI's tactical period games. *Grenadier* deals with the 1680–1850 span, covering the Napoleonic era, but Napoleonic enthusiasts may prefer one of the many games specializing in this theme.

GRUNT, SPI (not in polls) Rarely seen now, and never very popular, dealing with tactical combat in Vietnam in 1965.

GUERILLA, Maplay (not in polls) A large map and neat counters simulate the activities of a Ghurka battalion in Sarawak during the Indonesian confrontation with Malaysia. Secret victory conditions, ambushes, infiltration, hidden guerilla units, canoes, helicopters and the construction of chopper-pads in villages. Great attention to detail, but supply rules odd; movement uses a rather old-fashioned method. (cv)

HELMS DEEP, Fact and Fantasy (SPI 6·0, 119) Pleasant Tolkien-based game, though with a less professional air than the big companies' products. *Siege of Minas Tirith* is rather more sophisticated, and perhaps also has a stronger flavour of Tolkien.

HELMS DEEP, SPI (not in polls) Part of *The Ring Trilogy*, out in mid-1977.

HIGHWAY TO THE REICH, SPI (not in polls) Massive company-level game on Arnhem, with a map composed of four 22″ × 34″ sections in a long 'corridor'. Covers the full operation, from the first paratroop landings to the final evacuation (if the game goes historically, that is). Morale rules and anti-tank effectiveness are included.

HITLER'S LAST GAMBLE, Rand (SPI 5·8, 136) Said to be strongly unbalanced.

HOOKER AND LEE, SPI (6·7, 44) Most popular of the second *Blue and Grey* Quads; *See Fredericksburg*. As in *Antietam*, poor Union organization restricts the number of their units which can move each turn. South can send flanking units off-board to return later (the Jackson manœuvre). Short (nine turns), but with interesting dilemmas for both sides.

HUE, SDC (SPI 6·1, 109) One of SDC's relaunched Pouch series; a splendid tactical game by John Hill. Decimal dice combat system hinging on terrain defence values. The map shows the Citadel and suburbs of Hue, and the game accurately simulates the fierce actions fought in the streets. Communists lead at first with their attempts to capture armouries, but massive USMC/ARVN reinforcements soon drive them onto the defensive. Full of period 'feel'. Company level. (cv)

HURTGEN FOREST, SPI (not in polls) Part of the *West Wall* Quad. Rather boring Big Push by the American forces. (cv)

INVASION: AMERICA, SPI (SPI 6·5, 65) The idea here is that North America has become isolated and threatened by invasion from any or all of the Soviet bloc, the Chinese bloc, and a South American alliance. The hypothesis is fanciful, but it gives an excellent excuse for a modern war fought over a giant multi-coloured map of the continent (see Chapter 1). Different scenarios feature various stages of the 'campaign', so that you can concentrate on the problems of invasion, or on the battle for the interior. Despite large numbers of units on every side, the scenarios have fewer turns than in most complex games, so they can be played in 4–6 hours. Some problems with play balance, as the defence (realistically enough) can maintain almost permanent air superiority by basing their planes well inland. The total effect is colourful, complex, and naturally rather abstract. (*See* Chapter 1.)

ISLAND WAR, SPI (SPI 6·3, 93) A very mixed bag, and the least popular of the Quads in the poll, but with some excellent games. Notable for realistic artillery rules: artillery and mortars caused more Second World War casualties than any other class of weapon, yet fair treatment of them is rare. Here, artillery can be used for soak-offs, and to boost combat strengths, which complicates life for the attacker as the defender adds his last. Stacking limit one unit. Some ground support. *See Bloody Ridge*, *Leyte*, *Okinawa* and *Saipan*. (mw)

JENA-AUERSTADT, SPI (7·1, 7) Most popular game from the *Napoleon at War* Quad, based on the 1806 battle. See *La Grande Armee* for a more strategic treatment. Two battles are connected by a strategic-movement track. The Prussians are too fragile for good balance. (cv-np)

JERUSALEM, SPI (not in polls) Part of the *Modern Battles II* Quad, set in the 1967 war.

JERUSALEM, SDC (5·7, 145) The lowish rating is odd, as many people enthuse about this John Hill game. The map is small but attractive – the word 'cosy' is appropriate – and the pieces exotically varied, while the rules are highly inventive, as well as light-hearted at times, as when describing an Arab terrorist, El Kutub, planting bombs: 'If a "6" is rolled, something went wrong and El Kutub has joined Allah.' The game is set during the establishment of Israel (1948), and hinges on running convoys through to Jerusalem. Exciting and colourful, *Jerusalem* does have drawbacks as a game: the Arabs are unable to do much until the Arab Legion arrives, and then the outcome may hinge on a few crucial die-rolls as very powerful forces meet and one side is annihilated, a natural consequence of small scale combined with a bloodthirsty CRT. But it's *fun*. 3–5 hours.

JUTLAND, AH (SPI 6·4, 80/AH 6·17, 11) This is a remake of an earlier edition, improving a number of small points. The game is at least half-way to miniatures, with a small search board and the actions taking place off-board: you need a large (at *least* 4′ by 5′) playing area on which the big counters can be placed. The search procedure is well done, and numerous complex rules govern the tactical combat and engagement procedure. Six hours plus: plenty for naval buffs to get their teeth into, but a bit different from normal board wargames.

KAMPFPANZER, SPI (SPI 5·8, 136) Companion to *Desert War*, dealing with tactical combined arms combat at the start of the Second World War. Nine scenarios involving British, French, German, Japanese, Russian and Czech tanks. Simultaneous movement, modifications for different weapon types, entrenchments, overruns, and the controversial panic rule which randomly makes some units disobey orders. Comparable to *Panzerleader*, with more nations but fewer different types of unit; infantry in particular play a smaller role. The map is mostly clear terrain. Generally simpler but less challenging than *Panzerleader*, except for the simultaneity of movement. *Panzer '44* is a more advanced version, based on later on in the war.

KASSERINE, SPI (not in polls) Part of the *North Africa* Quad. Movement channelled into the gaps in a maze of rough and broken terrain, helping a difficult US delaying action against a powerful Axis assault. Less fluid than most North African campaigns, with tense struggles for key positions.

KASSERINE PASS, Conflict (SPI 6·4, 80) Fast-moving struggle in the mountains, with artillery playing a crucial role. Simple basic system makes it highly playable (a John Hill design); evenly balanced.

KHALKIN-GOL, SDC (SPI 5·5, 159) Simple system with considerable realism. The battle (from the Soviet-Japanese war) is fought amid the sandy plains and marshes of Mongolia. Mostly infantry armies with a few tanks, cavalry and artillery. A tough battle – including infantry leaping onto tanks to destroy them! Regimental level.

KINGMAKER, Philmar/AH (SPI 7·0, 15) Hexless and very different from the usual wargames in most respects, *Kingmaker* is a lively multi-player game loosely based on the Wars of the Roses. It was originally produced by Philmar, the British distributors of *Diplomacy*, and appeals to a similar audience. Emphatically not for those seeking a test of skill, the game becomes riotous with half a dozen players as nobles get whizzed home to look after local rebellions, plagues lay waste to armies, pretenders to the throne cower in Calais, and sittings of Parliament dish out high offices to powerful factions. The Philmar version is cheaper and has a bigger mapboard, but the AH version has better Parliament rules and other improvements, including some clarifications of obscure points.

KOREA, SPI (SPI 6·1, 109) This war should be excellent wargaming material, with its swift changes of fortune as first the West and then the Chinese entered, but for some reason it has been neglected apart from this game (see also *DMZ*). There are three scenarios featuring different stages, and a campaign game linking them together. Most games follow history closely, with the initial Northern invasion just failing to win before the western forces under the UN insignia drive them back deep into North Korea, whereupon the Chinese intervene and push back the UN, with a stalemate developing around the 38th parallel. There are rules for naval gunfire, sea transport, invasions, fortifications and supply, but neither side has a great choice of strategies and the general effect is not very challenging. A good simulation and a passable game. Moderate complexity, 3–4 hours per scenario or 10 hours for the campaign.

KRIEGSPIEL, AH (SPI 3·5, 202/AH 4·96, 25) Resoundingly relegated to bottom position by both polls, the game is far too simple for experienced wargamers. Unfortunately, it is also of doubtful value to beginners. There are plenty of interesting rules: invasions, weather, prisoners, supply, and an intriguing diplomacy rule, allowing negotiations which may have unexpected results. Tactical cards are used, as in *Caesar's Legions*, but here they determine combat rather than merely modify it, and no die is used. The trouble is that the game's scale is too small, with the result coming after a few brisk firefights before any plan can really get under way. Nor is it a good introduction to other wargames. 1–2 hours.

KURSK, SPI (SPI 5·3, 173) Divisional-level simulation of the German attempt to turn the Soviet tide after Stalingrad. The system in *Kursk* was later used in *France 1940* (originally an SPI game, now marketed by AH). Motorized units have a second movement phase, and there aircraft are included in an elementary way. *Kursk* did quite well in the polls after its appearance in 1971, but has slid since as more advanced simulation techniques for tank-dominated conflicts have been developed. Moderate complexity, 3–4 hours.

LEE MOVES NORTH, SPI (SPI 6·1, 109) Strategic-level version of the Antietam and Gettysburg campaigns (for tactical treatments of the actual battles, see the popular SPI Quad games *Antietam* and *Cemetery Hill*, and *Gettysburg* from AH), in which the Confederacy penetrated the Mason-Dixon line in 1862–3. Rules for limited intelligence and cavalry probes, supply, leadership, and four scenarios giving 'alternative history' possibilities. Map covers the relevant parts of Virginia, Maryland, and Pennsylvania in reasonable detail. Lee fans see also the next game, *Hooker v Lee*, *Battle of the Wilderness*, *Wilderness Campaign*, and *Chancellorsville*.

LEE *v* MEADE, Rand (SPI 4·8, 186) Low rating perhaps caused by the unfamiliar movement system, which appears again in *Omaha Beach*, rated 195th for no very obvious reason.

LEGION, SPI (SPI 6·7, 44) *See Chariot.* Connoisseurs of tactical Roman warfare can also consider *Caesar's Legions.*

LEIPZIG, SPI (SPI 5·5, 159) Strategical simulation of the campaign leading up to the dramatic Battle of Leipzig. Napoleon's army defeated at Moscow is menaced by separate Prussian, Russian and Austrian armies. There are seven scenarios and a campaign game, with rules for possible Austrian neutrality, defection of Napoleonic allies, and great emphasis given to leader counters; the strategic problems of long campaigns are covered with rules on supply, regrouping and attrition en route.

LEYTE, SPI (SPI 6·1, 109) Part of the *Island War* Quad. Slow, ponderous and predictable: first scenario is a Japanese strategic retreat westwards, the only drama being the fate of the southern garrison coping with an invasion on both flanks and only one road home; second scenario is a re-enactment of the First World War in mountainous terrain; third scenario simply has the Japanese being sat upon. (mw)

LIGNY, SPI (not in polls) Part of the *Napoleon's Last Battles* Quad. Potentially fairly long, unless the Prussian defence breaks quickly against Napoleon's set-piece assault. Tough fighting, with poor Prussian morale and French strength balanced by the powerful defensive positions.

LITTORIO, DCC–AWA (not in polls) Strategic South American fantasy: Brazil reaches for continental domination.

LITTLE BIG HORN, Tactical Studies Rules (not in polls) Comparable to *Custer's Last Stand*, with a very large map.

LORD OF THE RINGS, SPI (not in polls) Campaign game from *The Ring Trilogy*, out in mid-1977.

LOST BATTLES (SPI 4·5, 193) Ill-received in the poll (the second-lowest SPI game by a large margin), this deals with conflict at battalion/regiment level in an anonymous area of the Eastern Front during 1942–4. Highly complex, with units building up to large formations, different armour and non-armour combat values, engineers, supply dumps, command units and air strikes. 3–4 hours. *Panzerblitz* offers a good alternative on the same general theme.

LUETZEN, SPI (not in polls) Fine member of the *Thirty Years War* Quad, with a special thrill for Scandinavians (rarely appearing in wargames) as Gustav Adolphus fights the climactic battle for domination of Europe against the last Imperialist force in Germany. The main special features of *Luetzen* are fog, rolling over the battlefield at unpredictable intervals, a useful Imperialist force which comes to the rescue of its embattled and outnumbered colleagues, and a very curious rule on the death of Gustav: should this occur, it may shatter the Swedes' morale, or alternatively (as actually happened) make them, in the words of the rules, 'rally to a grim, all-consuming ferocity'. The latter outcome is most likely if they are doing badly at the time, on the backs-to-the-wall principle. The Imperialists gain a rich bonus for killing Gustav, so suicide is effectively ruled out.

LUFTWAFFE, AH (SPI 5·7, 144/AH 6·13, 12) This attractively-packaged game is unusual in that it features the American *strategic* air war against Germany, whereas most air games are tactical. Bombers with escorts stream in from different staging areas, and the German fighters fly off in waves to hit them before they reach their pre-planned targets. The major constraint on all fighters is fuel, and some of the best defending aircraft need constant landings for re-tanking. The

Basic Game represents one raid and is playable in a couple of hours; it is very satisfactory for beginners or anyone wanting a quick game, but rather too dependent on single die-rolls. The Tournament and Advanced versions are much longer, requiring up to ten separate raids, and have some additions to increase realism. Play balance is doubtful, with an apparent Allied edge in several versions, but using the right optional rules can cure this.

MACARTHUR, Research (not in polls) Brother of the disastrously rated *Patton*, by a company best known for non-wargames. Three battles of Macarthur on different boards are included. Very simple indeed.

MANASSAS, GDW (SPI 5·2, 176) This game was first published on an amateur basis by the designer, Tom Eller, and is now being produced in substantially the same form by GDW. The SPI poll is based on the amateur edition, which may have prejudiced the very small sample (one per cent of the total poll) who had seen the game against it. At all events, the game was highly praised in its amateur days, and GDW doesn't buy rubbish. *Manassas* is a brigade/regiment level simulation of the first battle of Bull Run, with 220 counters. Movement is simultaneous, and combat is by step reduction, with a slightly odd system in which each unit is accompanied by a marker showing current strength. There are night disengagement, weather, supply and military formation rules. Not for beginners; highly influenced by miniatures techniques.

MARCH ON INDIA, Jagdpanther (not in polls) Complex game covering the 1944 Imphal and Kohima battles. Accent heavily on supply (by road, rail or air). Battalion/brigade level. Long map shows the flank from Tiddim to Kohima with mountains and Manipur and Chindwin rivers. The Japanese hold the strategic initiative, and must endeavour to cut the roads before they run out of supplies. The British have a system of brigade boxes to harry the enemy. Fourteen turns; long, hard and exciting. (cv)

MARINE, Jagdpanther (SPI 5·0, 182) Forerunner of *Airborne*, but on marine operations. Nine scenarios include rescue of POWS, installation raids, and an army/marine exercise. The playing 'feel' is as described under *Airborne*. Neat tactical game with some very interesting subjects. (cv)

MARENGO, SPI (SPI 6·7, 44) Part of the *Napoleon at War* Quad, rated lower than the others. The Austrians seek to deploy swiftly to overrun the divided French army; the effect of cavalry is rather artificially simulated. (cv-np)

MECHWAR '77, SPI (SPI 6·7, 44) Brother of *Panzer '44*, dealing with tactical armoured combat in the present decade, in the Middle East, and hypothetically (we hope) between US and Soviet, and Soviet and Chinese forces. *MechWar '77* uses a recognizably similar system to its brother, but with various modern innovations: helicopters, smoke and ammunition depletion. Large, attractive map with imaginary German names (most of the scenarios are US-Soviet). Movement is sequential, but combat simultaneous, compromising between playability and realism. Fairly complex; 4–6 hours for average scenarios.

MIDWAY, AH (SPI 5·9, 128/AH 6·12, 13) This game is twelve years old, but still played by people interested in the context. The Japanese need to take Midway Island quickly, and each side also scores points for sinking enemy ships. The first part of the Japanese fleet has nearly all their aircraft, and is rather vulnerable to US attack on the first day. If they survive in good condition, massive reinforcements make the Japanese fleet a very tough nut to crack. Ships move hidden behind a screen and are sought by air and sea patrols; actual attacks are fought out on a tactical board. Surface combat is rare, which is just as well for the undergunned US in this battle. The game is often tense and exciting, though realism is limited and the Japanese have a definite edge in the usual game (this can be corrected by varying search capacities). (*See* Chapter 8.)

MINAS TIRITH, SPI (not in polls) Out in mid-1977. Part of *The Ring Trilogy*, on the siege of Minas Tirith.

MINUTEMAN, SPI (not in polls) Partisan warfare in America leading to a full-scale contemporary Second Revolution. The Minutemen are individual guerilla organizers, who attempt to build up groups of rebels in different parts of the country, against the efforts of counter-intelligence, informers, and regular army troops.

MISSILE BOAT, Rand (SPI 5·5, 159) Tactical warfare with modern naval techniques: electronics, missiles, advanced torpedoes, aircraft, submarines. Technically well-produced on miniatures lines, the game is interesting as a simulation but weak on skill; in particular, there is a silly version of a combat matrix in which, unlike that in, e.g., *Caesar's*

Legions, the interaction with one's opponent's choice is random, so one choice is as good as another. Mainly suitable for players with a special interest in current naval combat.

MISSILE CRISIS, DCC–AWA (not in polls) Hypothetical invasion of Cuba, 1962. Large map, varied colourful units (air/sea/land and submarine). Fairly simple; moderate length. The victory condition for the US is the destruction of the missile sites.

MISSILE PATROL BOAT, GDW (not in polls) Out in 1977: simulates small craft combat much as *SSN* deals with submarine combat.

MODERN BATTLES, SPI (SPI 6·8, 33) Quad composed of *Chinese Farm* and *Golan* (Arab-Israeli, 1973), *Mukden* (Sino-Soviet) and *Wurzburg* (Nato–Warsaw Pact). CRT complex for a Quad; numerous bloodless retreats. Artillery added to attack and defence (cf. *Island War*); air support; SAMs; fine maps. (cv-np)

MODERN BATTLES II, SPI (not in polls) This Quad will be out in mid-1977, with *Bundeswehr*, *DMZ*, *Jerusalem* and *Yugoslavia*, all except *Jerusalem* hypothetical.

MOSCOW CAMPAIGN, SPI (SPI 6·2, 102) September, 1941 saw the Germans within striking distance of Moscow, with all the momentum of their earlier victories. Stalin decided to keep the Government in the city and stand and fight, and the Germans just failed to break through the dense lines of defence, in one of the most crucial battles of the war. Promising stuff for a wargame, though necessarily something of a pitched battle. This one features automatic victories, separate CRTs for the two armies, the effect of five lines of defensive fortifications, and weather, with eight 'what-if' variations in forces. Three shortish (2–4 hours) scenarios cover October, November and December, and a campaign game links the three. Moderately complex, detailed map.

MUKDEN, SPI (SPI 6·6, 54) Least popular game in the poll of the *Modern Battles* Quad, perhaps because of its obscure location in a hypothetical war. Chinese militia can form guerilla bands to cut off the advancing Russians. (cv-np)

MUSKET AND PIKE, SPI (SPI 6·8, 33) *See Chariot. Musket and Pike* deals with 1550–1680, with firearms turning warfare into a quite different affair from the Agincourt era, but before cannon had completed the transformation. One of the most popular of the tactical period games.

NAPOLEON AT WAR, SPI (SPI 6·9, 23) Highly successful Quad comprising *Battle of Nations*, *Jena-Auerstadt*, *Marengo* and *Wagram*. The high rating in the poll is impressive, since the voters seem a little hard to please when it comes to the simpler Napoleonic games.

NAPOLEON AT WATERLOO, SPI (SPI 5·8, 136) Excellent introductory game (incidentally, hitherto free to new subscribers to *Strategy and Tactics*), exciting, easy to learn, and over in an hour (hence a favourite at one-day conventions). However, there are only fifty-plus units, so experienced players will not find it very challenging, and the game techniques are no longer all that good an introduction to recent designs. Artillery, demoralization and pinning of enemy units are covered.

NAPOLEON AT WATERLOO EXPANSION KIT (SPI 6·6, 52) Much more to the taste of the hard-core, but still at the easy end of the scale, this supplement to the main game includes a new set of units (brigades instead of divisions) and four pages of new rules. Good second game for those who started with the basic version; an alternative in the same category is *Borodino*. Game length rises to 1–2 hours.

NAPOLEON'S LAST BATTLES, SPI (not in polls) Quadrigame with a difference: the four constituent parts, *La Belle Alliance*, *Ligny*, *Quatre Bras* and *Wavre* combine to produce a game on the full campaign around Waterloo, as well as being available individually as usual. Simple rules resembling *Napoleon at Waterloo*, with demoralization, stacks of two and plenty of interesting terrain added on the highly attractive maps. Interesting command rules appear in the campaign game, with units separated from their leaders unable to make attacks.

NAPOLEON'S LAST CAMPAIGNS, Rand (SPI 5·5, 159) The 1814–15 period leading up to Waterloo. Another area (as opposed to hex) movement game; why this should be so much more common in pre-twentieth century simulations than in the *more* mobile modern wars is not immediately apparent, but may have its explanation in the short-ranged weaponry of the time. CRT modified by combat matrix as in *Caesar's Legions*; also leaders, fortresses, forced marches, supply and cavalry rules. There may be a bias against the French. The map is notably good.

NARVIK, GDW (SPI 6·9, 23) One of the Europa series but not yet combinable with the others, as explained under *Drang Nach Osten*; *Narvik* alone is played at regimental level, but divisional counters are provided for the Europa series. The German player holds the initiative throughout, but is hampered at every turn by Norwegian delaying actions, British carrier-borne air interventions, and a tight time schedule. Tense and well-balanced. There is a heavy emphasis on air combat and transportation lines through the difficult Norwegian terrain.

NATO, SPI (SPI 6·7, 44) The poll takes a jaundiced view of many operational-level modern games, but it rates this one well, and with good reason. The main battlefield is West Germany, after an invasion by the Warsaw Pact, who can choose to attack at once with the Western defences in disorder, or to build up the assault forces to a seething mass of units before crossing the border. The Nato forces reel back early on, severely hampered by the forces of the different nations each needing their own supplies. Things rapidly look very jolly for the attackers, but then things start to go wrong: the ferocious casualties the West can exact start to take their toll, the supply forces are left out of range by the forward troops, and Nato forces form a gradually stiffening line. Rules for tactical nuclear attacks and the extremely useful Western Tricap divisions, which can slip through a zone of control, also help the West, as does a curious rule on Denmark requiring a Soviet garrison if the Warsaw Pact absent-mindedly isolates it from Germany but not otherwise. Some queries on realism but thrilling from start to finish. 5–9 hours. (*See* Chapter 6.)

1918, SPI (SPI 5·9, 128) Germany's last fling in the First World War, with Stosstruppen adding savage punch to the assault on the Anglo/French line. The problem of First World War games is to simulate trench warfare without making things boring, as in the marvellously detailed *1914* (AH, but now out of print and only available secondhand), which was lovely to look at but a dour struggle indeed. *1918*, however, gives a good chance of German breakthroughs, the main constraints on a rapid advance being supply and the difficulty of getting adequate artillery support. 3–4 hours, medium complexity. *See World War I* for a good grand strategic treatment of the war.

NORAD, SDC (SPI 4·6, 192) May be out of print shortly, and with few adherents anyway, despite its handsome map. A strategic game of modern superpower confrontation.

NORDLINGEN, SPI (not in polls) The best of the *Thirty Years War* Quad, with strategic problems rearing their ugly heads beside the usual tactical questions in Quad games: a powerful Swedish force is threatening to overrun the Imperialist left flank, while the main body of Scandinavians waits nervously behind an imposing row of artillery pieces. If the Imperialists get their calculations right they probably have an edge, but both sides have nail-biting dilemmas throughout. (*See* Chapter 7 and Part v.)

NORMANDY, SPI (SPI 5·9, 128) Regimental-level simulation of the struggle for the beaches up to D-Day plus six (see *Breakout and Pursuit* for the campaign afterwards). German defences are unknown to the Allies as they storm ashore, and six German orders of battle are given to cover different historical possibilities. Rules for naval gunfire, paratroops, commandos and supply; plenty of units but a mere six turns, so playable in 3–4 hours. *Overlord* is an alternative, and *D-Day* simulates the whole campaign to the Rhine at a more strategic level.

NORTH AFRICA, SPI (not in polls) Quadrigame distinctly more complex than usual, though still very playable. Artillery rules of the *Island War* type, CRT columns varying with the terrain as well as the odds, and detailed terrain variations on attractive maps. The components are *Cauldron*, *Crusader*, *Kasserine* and *Supercharge.*

THE OCTOBER WAR, SPI (not in polls) Out in spring 1977, with tactical armoured combat in the Middle East, during the 1973 war. Platoon/company level on Golan and Sinai terrain. See *Modern Battles* for operational-level treatments.

OIL WAR, SPI (SPI 5·3, 173) During the Arab oil boycott in 1973, there was a good deal of speculation on the feasibility of an American intervention to seize the wells on the grounds of economic self-protection. This game simulates this possibility, together with some rather less likely possibilities such as an Iranian assault on Iraq with US intervention on the Iraqi side. The game system is not very complex, but unusual, with the entire US force air-lifting into the action. The US has air domination, but the defenders have considerable ground superiority, and the race to seize the wells in the eight turns allowed is touch and go. Some interesting problems for both sides, but probably not enough sustained suspense for the hard-core. Playable in a few hours.

OKINAWA, SPI (SPI 6·4, 80) Part of the *Island War* Quad, with a massive sixty turns, unusual in a Quad game. (mw)

OMAHA BEACH, Rand (SPI 4·4, 195) Unusual movement system: squares instead of hexes, with movement costs printed around the edges. Highly tactical, with mines, strongpoints, a rough sea landing, and a pitched battle on the beach. *See Normandy*, *Overlord*, and *D-Day* for progressively more strategic treatments. Very considerable rule problems and omissions: errata sheet essential.

OPERATION OLYMPIC, SPI (SPI 6·0, 119) The mediocre poll rating almost certainly reflects an aversion to solitaire games, as this is widely believed to be the best of the type. The only wargame fought on Japanese mainland soil, it poses the problem of storming the Imperial heartland in the absence of an atomic bomb. Solitaire games are especially suited to players who don't know other nuts, but on the one hand most games can be played solitaire (except hidden/simultaneous/multi-player types), and on the other hand it is possible to run postal contests in solitaire games, with each contestant getting the same die-rolls, though this is rare at present. *Olympic* is fairly complex, but realism yields to playability where necessary.

ORIGINS OF WORLD WAR II, AH (SPI 5·3, 173/AH 6·02, 16) Unhappy attempt to marry wargames with multi-player diplomacy: the game is quite interesting, but marred by violent bias to certain countries. Britain, France, Germany, the USSR and the USA jostle for popularity in the other countries by committing political factors to them and having them 'fight for influence' with a wargames-type CRT. Germany and the USSR dominate the historical scenario, with no American victory ever having been recorded to my knowledge, though the rules claim it can be done. Sometimes played by post in Diplomacy magazines. Often tense, and easy to learn. Play length 1–2 hours, five players essential.

OUTREACH, SPI (not in polls) Successor to the popular *Star Force*, with supposed developments of the space technology in that game. Rival civilizations race to explore, settle and dominate the Galaxy. Movement is by 'stellar shift', with the range unlimited but the likelihood of scattering increasing with distance. Civilization levels, autonomous and partially uncontrollable alien forces, investment in colonization and both military and peaceful ships, and various plausible scenarios feature, with an entertaining diplomacy rule whereby both players benefit if both vote for cooperation, but each may stand to gain from aggression, either as a surprise stab or a pre-emptive strike – a situation known to game theorists as Prisoner's Dilemma, and tending to encourage treachery.

OVERLORD, Conflict (SPI 6·1, 109) Normandy landings game by the makers of *Kasserine Pass* and *Bar-Lev*.

PANZER ARMEE AFRIKA, SPI (SPI 6·6, 54) Enduringly popular operational-level game of North African campaign. Lightning movement emphasized with movement factors up to sixty, detailed supply rules, and the controversial Command Control system, whereby certain randomly-selected units on the Allied side fail to get their orders (also used in *Sniper* and other games), simulating real-life confusion ... realistic and extremely irritating for the player affected! Moderate complexity, good choice of strategic and tactical options, rather realistic. Not very suitable for postal play. *See Afrika Korps* for a simpler but less realistic game, and *Tobruk* and *Desert War* for tactical-level combat on this front.

PANZERBLITZ, AH (SPI 6·8, 33/AH 6·42, 7) Perhaps the most frequently played wargame ever produced, *Panzerblitz* was the first to bring a wealth of tactical detail to the Second World War East Front, and met a delighted reception from the hobby when it came out in 1970. Although game design has moved on since, it still has many adherents, and anyone interested in tactical armour/infantry warfare should try it. Three boards from anonymous sections of the Soviet countryside feature a rich variety of hills, riverbeds, villages, woods and winding roads, and can be fitted together in various ways to make different maps. Units have four factors (attack, defence, range, movement), and combat is modified by armour and weapon type. Drawbacks are somewhat unbalanced scenarios and the 'Panzerbush' syndrome, in which units popping from wood to wood cannot be attacked by non-adjacent units, which is a flaw in realism. Exciting, high skill level, very complex; 2–4 hours, depending on scenario. *See Panzerleader* for companion game and *Kampfpanzer* for SPI's simultaneous movement alternative. *See* Chapter 7 and Appendix C.

PANZER 44, SPI (SPI 6·8, 33) Second World War equivalent of *MechWar '77*, featuring tactical armoured warfare on the West Front in 1944–5. *See MechWar* and also *Kampfpanzer*.

PANZERGRUPPE GUDERIAN, SPI (not in polls) The battle for Smolensk. Fluid operational game with 'attempted overruns' allowed: if the attack succeeds, you can carry on; if not, you must stop. Soviet organizational units, supply, untried forces of unknown strength feature, together with air and partisan interdiction.

PANZERLEADER, AH (SPI 6·8, 33/AH 6·5, 3) Produced four years after *Panzerblitz*, this West Front game is more of a son than a brother, as the game system is similar but with certain distinctive new features. The most important ones are that the 'Panzerbush' tactic is abolished, with anyone trying it liable to be transfixed by 'opportunity fire' on the way (this makes postal play more difficult), and the hexes have spots in the middle to facilitate calculation of lines of fire from one to another (to see if some damned hill is in the way), which is useful. This time there are four mapboards, one of them a beach to allow for landings. The opportunity fire rule, while more realistic, has a slight tendency to keep units' heads down in cover, so the game is not quite as fluid as *Panzerblitz*. However, both games are excellent, and preference is largely a matter of taste. It should be noted that neither uses the ultra-detailed miniatures technique of distinguishing between different types of hits (on a turret, or tracks, for instance), unlike e.g. *Tobruk* and the Battleline tactical games; the miniatures approach is more realistic but slows things up with extra die-rolls, as well as adding more random factors.

PATROL, SPI (SPI 6·8, 33) A companion game to *Sniper*, *Patrol* deals with individual combat from the First World War up to the present. Brisk scenarios, with the flavour of man-to-man fighting quite well reflected, as the players agonize over whether to try and pin the enemy down or make a rush for it, whether to concentrate the squad or spread them out, and over the possible enemy plans.

PATTON, Research (SPI 3·7, 201) Second only to *Kriegspiel* in unpopularity in the poll, with a full half-point to the two hundredth placed game. Three battles, each on its own board. *See Macarthur.*

PEARL HARBOR, GDW (not in polls) Out in early 1977. Designed by John Prados of *Third Reich* fame and covering the whole Pacific War, from 1939 to 1945. *Cf Coral Sea.*

PORT ARTHUR, GDW (not in polls) Mated with *Tsushima*, the naval aspect of the Russo-Japanese war, this gives the land struggle. Troops arrive in the area from Japan and Europe by strategic movement chart. The crucial struggle is for the port itself, the only warm-water one available to the Russians in the area; this makes the joint game with *Tsushima* particularly interesting, though *Port Arthur* can be played on its own with abstract naval rules. Distinct rule problems on supply and stacking; ask for an errata sheet. Interesting to see Mukden appear here in a major role as well as in the *Modern Battles* Quad. An exciting place to live, evidently.

PURSUIT OF THE BISMARCK, DCC–AWA (not in polls) Scheduled for 1977 release: like *Jutland*, this is basically miniature-oriented, with 1:1200 scale counters. Naval/air game emphasizing the *Bismarck*'s hunt for convoys rather than the final chase to destruction. Note that an AH game on the *Bismarck* is expected out as well, with a simple basic game and a highly detailed advanced game in the style of *Tobruk*.

PUNIC WARS, SPI (SPI 5·4, 169) Area-movement game on the three wars between Rome and Carthage between 260 and 201 BC. One of history's more important conflicts (Cato used to start every speech with a thunderous 'CARTHAGE MUST BE DESTROYED'), but the game balance is dubious, especially in the third war.

QUATRE BRAS, SPI (not in polls) Part of the *Napoleon's Last Battles* Quad, with a gripping battle for the crossroads which swings dramatically to the French and back again.

QUEBEC 1759, Gamma Two (SPI 6·0, 119) Drastically different from usual designs, this game has a long, attractive map of the Heights of Abraham, curious domino-like wooden blocks as units, and simultaneous area-movement; step reduction, logistics and naval units feature. Makes an interesting change but out of the mainstream of board wargames.

RAIDERS OF THE NORTH, DCC–AWA (SPI 4·9, 185) Sister game to *Battle of the Atlantic*, and again criticized for physical quality. Scenarios on the major Second World War Atlantic naval engagements. Large search board; combat on separate battle board. Rules for torpedoes, destroyers, radar, submarines and fire control.

RED STAR/WHITE STAR, SPI (SPI 6·2, 102) Ten-scenario game of tactical battles in Southern Germany in a hypothetical contemporary war, with platoon, company and battalion-level US, West German and Soviet counters. Wire-guided anti-tank missiles, rocket launchers and helicopter gunships have starring roles; the total effect is highly complex (see *MechWar '77* for a rather more elementary, highly playable alternative). 2–3 hours per scenario, once the rules have been absorbed.

REMAGEN, SPI (not in polls) Part of the *West Wall* Quad. A rather contrived simulation of the capture of the famous bridge, and the ensuing battle. (cv)

REVOLT IN THE EAST, SPI (not in polls) Lively corps/army level simulation of anti-Communist revolts in Eastern Europe, aided by Nato intervention; the Soviet units have the advantage in stand-up fights, but are hard put to cover each rebellion as it breaks out. Distinctly odd political assumptions, e.g. Bulgaria almost as likely to revolt as Rumania, but fun as a game, though rather strongly luck-dependent.

RICHTHOFEN'S WAR, AH (SPI 6·7, 44/AH 6·48, 4) One of the relatively few pre-1974 games in SPI's top fifty, this game has surprised many with its continuing success with the rarely-simulated theme of the First World War tactical air combat. Of its kind, it is an excellent product: clear rules, well balanced between playability and realism, a variety of brisk scenarios, and good period 'feel', with a more interesting mapboard than usual in air games (showing ground targets), avoiding the usual temptation to breathe a sigh of relief and produce a blank sheet with a few cloud counters. Average scenario length an hour, with dogfights, trench-strafing, photo-reconnaissance, balloon-bursting and bombing.

RIFLE AND MUSKET, SDC (SPI 4·7, 189) Hard to obtain now. In general, this sort of tactical skirmishing is ill-favoured by the hardcore, which helps to explain the low poll rating of this game, and *Rifle and Sabre.*

RIFLE AND SABRE, SPI (SPI 5·5, 159) See *Chariot.* The usual tactical palefaced map, with the open terrain modified a little more than usual, however, by a large clump of hills. Numerous units to simulate engagements from the American Civil War, Franco-Prussian War, Boer Wars, Spanish-American War, and many other conflicts. Based on but markedly simpler than *Grenadier*; artillery, mounted rifles, cavalry, primitive machine-guns, morale, shock tactics, and entrenchment rules. 2–3 hours.

THE RING TRILOGY, SPI (not in polls) Out in May/June 1977, with two battle games on Minas Tirith and Helms Deep, plus a two-map campaign game, *Lord of the Rings*, based on the Tolkien trilogy.

ROAD TO RICHMOND, SPI (not in polls) The crucial three days of the 'Seven Days Campaign' between McClellan and Lee. An early Confederate advantage is gradually offset by Union reinforcements.

Special features are a small Command Control Zone replacing the limited-movement rule in *Antietam*, a Union train unit, and Union Siege Artillery. Out in early 1977.

ROAD TO RUIN, SPI (not in polls) Out in August 1977. The 1942 Axis Summer Offensive, from Voronezh to Stalingrad, by two German armies, with untried units and shifting objectives for the German player.

ROCROI, SPI (not in polls) Part of the *Thirty Years War* Quad, and possibly the only wargame to date designed by a woman. The map of the battlefield between the French and Spanish armies is mostly blank, but the markedly different movement and combat strengths of the units on each side give each side interesting tactical problems; if the Spanish and Walloon regiments in the centre can bring the enemy to battle early on, they are likely to carry the day, but the French have a fair chance of keeping away while they chew the Spanish flanks. The optional leader rule does not work clearly and should be skipped. (*See* Chapter 2.)

ROMMEL, Loren Sperry (SPI 4·7, 186) The low rating is probably due to the semi-amateur flavour given the game by its unmounted board and pieces. In fact, it is an interesting development of the techniques of *Afrika Korps*, using step reduction and breakthroughs à la *Anzio* instead of the automatic victories at 7–1 and 'sudden death' CRT of *Afrika Korps*. The campaign game stretches an extra nine months, and there are three turns a month instead of two, so the total length is 15–25 hours; however, mini-scenarios lasting a few hours are given. *Afrika Korps* has more blood and thunder, but *Rommel* is probably more realistic and less luck-dependent.

ROMMEL'S WAR IN NORTH AFRICA, Rand (SPI 5·5, 159)

THE RUSSIAN CAMPAIGN, AH (SPI 5·9, 128) The SPI poll is from the game's previous incarnation as a Jedko game. A corps-level simulation of the Soviet front in the Second World War, notable for lots of units and a bloodthirsty CRT. Hitler and Stalin have their own counters, and the general effect is a lively 'fun' game rather than a deadly serious study of the war. Rather more complex and detailed than *Stalingrad*, AH's alternative, but less so than *Barbarossa*, SPI's game on the subject, which in turn pales in comparison with *Drang Nach Osten*. It tells us something about the hard-core voters that they put the most complex game in first place, with the others following in order of difficulty at 119th, 128th, and 150th. It is a natural process to want a greater challenge after one has mastered a simple game, and there is something to be said for buying increasingly complex games on the same subject, as one then relishes the exotic innovations each time. *The Russian Campaign* is 4–8 hours.

SAIPAN, SPI (SPI 6·4, 80) Part of the Island War Quad. Tense battle for the Island; each side must wipe out the other to win. Bloodless CRT, with units tending to retreat until the sea is at their backs and they have to stand or die. Japanese 'Banzai' attacks boost their initial attempts to throw back the US invasion; failing this, they retreat step by step into the northern tip and try to hang on until game end. Historically inaccurate and hard for Japanese to do better than draw, but exciting and often cliff-hanging to the end. 3–5 hours. (mw)

SARATOGA, 1777, Rand (SPI 4·3, 198) Oddly enough, this appears to be the only simulation of the Saratogan battle, with the American Revolution little covered from the tactical viewpoint; the relatively large 6% of the poll who have the game, however, put it near the bottom.

SCHUTZTRUPPE, Jim Bumpas (SPI 6·3, 93) Voted one of best three amateur games, 1976, in poll of manufacturers and magazine editors. Deals with German guerilla campaign in East Africa, World War I.

SEARCH AND DESTROY, SPI (SPI 5·8, 136) Tactical Vietnam combat; powerful air-mobile forces combated by hidden NLF guerillas. Unsuitable for solo play because of the hidden movement.

SEELÖWE, SPI (SPI 6·2, 102). SPI are too modest about this game; the rules go to great lengths to explain how far-fetched the historical assumptions had to be to allow this hypothetical German invasion of Britain, 1940, to go ahead, but a good case can be made that the defeat of the RAF would have made it impossible for the Royal Navy to intervene effectively, the main assumptions of the game. Players of the game should read *Sea Lion* (Richard Cox, from Futura Publications, 110 Warner Road, Camberwell, London SE5), a fictionalized book based on a wargame on the operation carried out by British and West German military officers in 1973. The SPI game is a thrilling struggle on the south (or in one scenario the east) coast, with German troops surging inland during good weather turns, with massive air support, and the British fighting back grimly whenever the clouds thicken to frustrate the Luftwaffe and the Channel turns rough, disrupting the lifeline of invasion supplies. Much depends on the luck of the weather, and the geography of the map is a little odd in places, but the excitement level is well above average. 3–5 hours. (*See* Chapter 4.)

1776, AH (SPI 6·6, 51/AH 6·44, 5) Strongly preferred to its SPI rival, *American Revolution*, in the SPI poll, *1776* is certainly very much more complicated. A lush board, and a tremendous range of counters showing the polyglot forces on both sides during the Revolution, predispose one to its favour from the start, but it is highly complex, filled with unusual rules, and the full campaign can take over ten hours. A fine game requiring time, enthusiam and concentration; definitely not for beginners. Some doubt about play balance; if the players find the British winning too often, they should adjust the rules to cut down such unhistorical delights.

SEVENTH CAVALRY, DCC–AWA (not in polls) Tactical squad level battles of Indian warfare. Limited step reduction; nine historical battles and a number of 'what if' scenarios. Good map, but very simple. *See Custer's Last Stand* for a good, though less tactical, alternative; there is also a big-map game from Tactical Studies Rules (of *Dungeons and Dragons* fame) called *Little Big Horn* which I understand to be rather similar to *Custer's Last Stand.*

SHENANDOAH, Battleline (SPI 5·2, 176) Not so much a wargame, more a way of life! Nineteen scenarios detail every possible aspect of the Civil War campaign in the Shenandoah valley, 1862 and 1864, ranging from the pretty complex Basic Game through the extremely complicated Advanced Game to the still more esoteric possibilities of the optional rules. The map is second only to *Anzio* in bewildering variety of terrain features; the 375 counters show infantry, cavalry, leaders, artillery, horse artillery, supply, forts and devastation; the rules cover different types of formation, complex stacking effects, detailed tactical combat, step reduction, garrisons, weather, forced marches, hidden Confederate formations, partisans and cavalry raids, with the B&O rail line playing a major role in the scenarios. Very hard work for the players, but rewarding for those interested in the campaign. Copious background and play notes. Scenarios range from a few turns to the entire campaign.

SHILOH, SPI (SPI 7·0, 15) Part of the popular *Blue and Grey* (I) Quad, but fought in very rough terrain, which is rather limiting. (cv)

SICILY, Rand (SPI 4·7, 189) Another of the less popular of the Rand package, despite the fact that this seems to be the only game about the invasion. Great variety of scenario possibilities, but various rule queries.

SIEGE, Fact and Fantasy (SPI 6·0, 119) Designed by Lou Zocchi; tactical sieges and storming of castles, with rams, peasants, knights, longbows, crossbows, onagers, fighting towers, and boiling oil; scenarios feature Hadrian's Wall, an AD 400 castle, and an AD 1400 walled city.

SIEGE OF JERUSALEM, AD 70 Historical Perspectives (not in polls). Tactical simulation of the Roman sieges during the Jewish Revolt in AD 66–72. Four cardboard map sections; five scenarios with a rule booklet each. Infantry, cavalry, siege towers, battering rams, catapults, onagers, zealots, leaders and special scenarios rules feature. Moderate complexity, fairly long.

SIEGE OF LENINGRAD, Jagdpanther (not in polls) Brigade/corps simulation of operations of Army Group North. Map shows front from Lake Peipus to Tikhvin. Five scenarios from the advance to the final defeat of the Army Group. Combat hinges on intelligent manœuvre as units may not stack adjacent to stronger units. Supply for Leningrad is handled at length to simulate the siege. Luftwaffe, winter, the Ice Road, and sensible supply rules. A good, clean east front game, about the most dramatic siege of the war, incidentally the only one to generate a symphony (by Shostakovich), played for the first time by radio broadcasting from the embattled city. (cv-np)

SIEGE OF MINAS TIRITH, Fact and Fantasy (not in polls) Includes a free mini-game, *Battle of the Slag Hills*. Just sixty counters, but Kings, Princes, and Magicians appear in the four-day siege.

SINAI, SPI (SPI 7·0, 15) The oldest (1973 vintage) game still in the top twenty, and a fine operational-level simulation of the 1948, 1967, and 1973 Mid-East wars, on the Syrian and Jordanian fronts as well as Sinai. Israel wins the first two wars comfortably (the game victory depending on the level of Israeli success), but in 1973, the better Arab training and SAM missiles make it very much of an even struggle, with a fascinating choice of strategies for the Israeli player in particular: he can defeat each opponent individually if he can afford to leave token forces against the others, but this can lead to disaster if not done with great care. Bloodless CRT, with manœuvre counting most in the Sinai;

armoured breakthroughs are immensely valuable if they can be achieved. 5–8 hours.

SIXTH FLEET, SPI (SPI 6·4, 80) Battle for control of the Aegean and Eastern Mediterranean, between Nato and Warsaw Pact naval and air forces. Fairly complex, with unusual game techniques: experienced players often find it hard to adjust to combat coming before movement, and there is a bizarre opportunity for submarines to avoid attack by retreating before combat, so that the only way to catch them is to surround them. Despite these possible drawbacks, and the gaps in realism caused by the changes in influence patterns around the Mediterranean since the game came out, *Sixth Fleet* is both interesting and exciting, with both sides racing against time as the Soviet forces strive to sweep the Aegean before the big US carriers slip past the Soviet submarines and reach the scene. The air rules are easy to operate and work well, with suitably awful results for players who send up too much at one time and have to refuel them all on the next turn. 4–7 hours.

SNIPER, SPI (SPI 6·8, 33) House-to-house fighting in the Second World War, with a counter for each individual – you can't get more tactical than that! Varying weaponry, multi-storey buildings, a wide variety of options for each man, and controversial panic (command control) rules frustrating your best-laid plans. Tanks, trucks and half-tracks in some scenarios, the tanks, unusually, being cardboard models which the players put together. Simultaneous movement. Exciting and fast-moving; lengthy rules, but easy to play once you have tried a game or two. Most scenarios playable in an hour or two. Companion to *Patrol*.

SOLOMONS CAMPAIGN, SPI (SPI 5·9, 128) Unusual design based on the campaign around Guadalcanal: just 200 counters and 23 giant zones; the simultaneous movement system enables each player to distribute his air, sea and land forces among the zones as he thinks most likely to frustrate the enemy. Strongly strategic emphasis, with naval task forces, combined arms operations and the problems of Japanese supply on Guadalcanal. 3–4 hours, quite complex.

SORCERER, SPI (SPI 6·3, 93) Produced in response to the boom in fantasy games, this is a hexed wargame with highly unconventional forces. 2–6 players take the part of Sorcerers specializing in certain colours; the board is multi-coloured, and the strength of the sorcerer depends on where he is standing. The forces of the contestants consist of magic bolts, vortexes, enchanted fortresses, air dragons, trolls, demonic infantry and boring old humans. Various scenarios with detailed pseudo-historical backgrounds, though not all are well-balanced. The counters are splendidly varied in numerous colours, and the game is refreshingly different; best for face-to-face rather than postal play. Special solitaire scenarios are included. 3–5 hours, complex with rather abstract feel.

SOUTH AFRICA, SPI (not in polls) Out in mid-1977, simulating full-scale warfare between Soviet-backed African nations and the white regimes in Rhodesia and South Africa. The game seems likely to be partially overtaken by events and should be highly topical.

SPANISH CIVIL WAR, Jagdpanther (SPI 5·6, 150) With *Battle for Madrid*, Jagdpanther have made a specialist corner for themselves in this prelude to World War II. The units are anonymous strength points, but differentiation is made between Nationalists, Basques, Anarchists, Republicans, Moroccans, Allied, Italian, German and International troops. Rules include armour, anti-tank and flak, with supply problems forcing the Republicans to cling to their cities. Three-player version separates Anarchists as third party. A close game with fairly simple rules. (cv-np)

SPARTAN, SPI (SPI 6·9, 23) *See Chariot. Spartan* scores best of the series in the poll, with battles from Marathon to Heraclea (where Pyrrhus won his defeat-in-victory). Main special feature is elephants, who have the nasty habit of running amok after being in combat.

SPITFIRE, SPI (SPI 5·8, 136) Completely blank map, but varied counters showing all the major aircraft which fought over Britain and France in 1939–42. Altitude is handled with counters for different levels, and the current actions of each plane are recorded on separate sheets, which have the aircraft specifications as well. Varying turning ability, fuel injection and some special manœuvres feature. *See Air Force* and *Basic Air Combat* for more complex treatments, and *Their Finest Hour* for a strategic approach. *Spitfire*'s simple rules facilitate the fast play which air combat enthusiasts tend to prefer, at the loss of some realism. Game length up to 1 hour or so.

SQUAD LEADER, AH (not in polls) Due out in 1977, this deals with single officers, NCOs and squads in the Second World War (*cf Sniper*) with techniques based on miniatures and *Panzerblitz*, and a simple game system. The designer, John Hill, is noted for colourful games, e.g. *Jerusalem.*

SSN, GDW (SPI 6·5, 65) Another featureless map, understandably as the theme is submarine warfare. Movement is simultaneous (and secret for submarines), as is combat. Probability of detection is calculated as a factor of noisiness and level of detection apparatus of each vessel. Weapons include all the major known current systems, and there is a wide range of scenarios and forces. *See Submarine* for a Second World War game on the subject.

STALINGRAD, AH (SPI 5·6, 150/AH 5·56, 22) When I first started wargaming in 1967, *Stalingrad* was the most challenging game around, with most players having their own pet strategies and theories. Designs have come a long way since then, and it is generally felt that *Stalingrad* is much too unrealistic in its details (large numbers of units have almost identical strengths on the German side, and the sudden-death CRT makes luck a major factor; moreover, the map is very short on detail). Even in the 'classic' range to which it belongs, other games, e.g. *Afrika Korps*, tend to be preferred. But while the tactical accuracy is faulty, the game gives a good simulation of the grand strategic alternatives, and it is swift and exciting, with more forward planning needed than in many games because of the savage effect of winter on movement. Fairly simple rules; Russians should win even in the 'balanced' version given even play and fair dice, but in practice things tend to even up. 3–6 hours. (*See* Chapter 3.)

STARFORCE, SPI (SPI 6·8, 33) One of the most successful and widely played space games, featuring grand tactical combat between humans and aliens across the galaxy. See *Outreach* for a sequel.

STAR RAIDER, DCC–AWA (not in polls) Vivid map with variable star configurations, but design shows signs of being first effort. Seven ship, three ground army and two missile battery types, with 300 counters in all. Violation of neutrals, ship capture, mining, secret weapons, third dimension and limited reconnaissance rules; three scenarios of varying length and complexity.

STARSHIP TROOPERS, AH (not in polls) This game is based on Robert Heinlein's science fiction classic, and should please anyone who enjoyed the book; others can also enjoy it, but may be a little bewildered at times! Vast variety of combatants (terrans, Skinnies and Bugs of various types) and equipment (proximity and remote bombs, nerve gas, listening devices and demolition charges appear). The rules are filtered in gradually through the six scenarios, with advanced versions of the early scenarios once all the rules have been absorbed; this is an advantage for beginners, but experienced players may be tempted to play the early version until they feel at home and then move on. The tactical problems are excellently done and true to the book, with powerful Mobile Infantry fighting numerous enemies hidden in tunnel complexes.

STAR SOLDIER, SPI (not in polls) 'Tactical combat in the twenty-fifth century', featuring the future history of *Starforce* at an individual combat level.

STELLAR CONQUEST, Metagaming (SPI 6·3, 93) Described in *Moves* by Richard Berg of SPI as 'far and away the best sci-fi game on the market. The basic game is simple, yet effective, and the advanced game and optional rules are quite intelligent and intriguing'. Inter-stellar colonization; stars with planets of unknown value; various levels of increasing production (and difficulty).

STRATEGY I, SPI (SPI 6·4, 80) Immensely ambitious project to simulate warfare throughout history with the same range of 1000 plus counters and two abstract maps, plus a vast choice of rule modules for every occasion. If you want to simulate a neo-colonial modern war, for instance, you look up the scenario rules, which tell you which modules and units to use and how to fit the maps together. There is a two-player Alexander-Darius scenario, but the others are basically multi-player games. The complex production rules of the modern scenarios are very alluring to the hard-core, but even the simpler wars of earlier periods are quite a handful. The main drawback, apart from the length (four hours to thirty or more), is a rather abstract flavour, so that one country is much like another. An impressive attempt, with much to interest would-be designers.

SUBMARINE, Battleline (not in polls) Everything said about *Air Force* applies here, in the different context of Second World War submarine warfare. Esoteric variations in armaments on both sides, and details of a long list of surface escorts and submarines on both sides and each front of the war. Admirable accuracy; complex; 1 hour upwards, with some lengthy scenarios.

SUPERCHARGE, SPI (not in polls) Part of the *North Africa* Quad, with three scenarios around El Alamein and Alam Halfa, including the Alamein battle itself. More open country than the other games in the Quad, apart from the main German defence line.

TACTICS II, AH (SPI 4·4, 195/AH 5·49, 23) Respectable but dull abstract introductory game, from the classic era. Better than *Kriegspiel* as a lead-in to other games, but readers of this book looking for an easy starting game would probably enjoy a simulation of an actual battle more.

TANK, SPI (SPI 6·2, 102) Armoured combat between individual tanks from the 1930s to the present era. Solid tactical game; there is also an expanded version for the ambitious.

TERRIBLE SWIFT SWORD, SPI (not in polls) Massive simulation of Gettysburg, with three giant and heavily-featured maps, and two sheets of counters at regimental level. Strongly tactical in emphasis, with particularly detailed rules for firing with different types of weapon. Rules for ammunition, leaders, complex combat sequence, and capture of enemies, plus numerous charts and tables. A magnificent simulation, but far from swift – however, as in *La Bataille de la Moskowa*, whose 'feel' it resembles, there are four small scenarios, giving the chance to try out various parts of the game before tackling the grand design as a whole.

THEIR FINEST HOUR, GDW (not in polls) Three games in one. A squadron-level simulation of the Battle of Britain develops the air rules used in *Drang Nach Osten*, with provisions for interception, specialized air rules, and losses varying with the control of the territory below. A group-level version is designed for slotting into the Europa series (see *Drang Nach Osten*). Finally, there is an invasion game, which can be played as a sequel to the air game if the Germans won, or as a separate game. The invasion game is at regiment/brigade level, with naval units at ship/flotilla level; land, sea and submarine combat appear, with air support rules. This game has a separate large-scale South coast map to the two-sheet map used for the air games, and it is the best.

THIRD REICH, AH (SPI 6·8, 33/AH 6·43, 6) The same theme as the Europa series, Second World War in Europe, but sufficiently playable to make it possible to get a reasonable way during one (preferably long) evening; the alternative game of the type is *World War II*, which is commonly thought a bit simpler but less challenging. *Third Reich*'s strongest point is the military production system, which works smoothly (if with questionable historical accuracy in certain cases) and ties the game together. Air, sea and land combat rages over the continent, and the slightest slip can lead to disaster with armour breakthrough and paratroop rules making every front insecure. Excellent both for two players or with more, though the rules tie the players closely to history and limit diplomatic opportunities. Extremely complex, with rule disputes not uncommon; every participant should be *good* to make the game really come to life. 6–10 hours, though with some mini-scenarios. (*See* Chapter 4.)

THIRTY YEARS WAR, SPI (not in polls) Quadrigame, made up of the excellent *Nordlingen* and *Lützen*, the good *Rocroi*, and the awful *Freiburg*. *Thirty Years War* has the virtues of nearly all the Quads: easy rules, rapid games and lots of combat with little messing about with staff work; conversely, the detailed realism of the really massive games puts the Quads in the shade. This Quad has particularly attractive artillery rules: guns are almost useless at a distance, but extremely effective in breaking up attacks at point-blank range. Should the emplacements be overrun, the counters can be turned over to reveal guns of the opposite colour – very neat. Leaders have a major influence, especially in rallying disrupted units, but if casualties reach a certain point the men sensibly stop listening to impassioned speeches and decline to go back on the attack. Worth getting despite *Freiburg*, and perhaps a bit of home redesign or an errata sheet can make this more satisfactory as well.

TOBRUK, AH (SPI 6·7, 44/AH 5·90, 18) The only case of an AH game which fares much better in the SPI poll! An extremely innovative simulation of North African tactical warfare, strongly influenced by miniatures concepts. The map is devoid of features, one grain of sand looking much like another, but there is a rich variety of units, both armoured and infantry, with various forms of static defences (e.g. mines) to put on board. Early scenarios are very simple indeed and not very interesting; later ones add more rules steadily, until the final scenario plays with the full orchestra of concepts, and the early scenarios have the advanced rules added on (see *Starship Troopers* for remarks on this approach). The most controversial feature is the legions of die-rolls required, as each round of fire is checked in exhaustive detail for chance of hitting and place of damage. Impressively detailed, with strong flavour of realism, but some miss the blood and thunder of faster-moving games. Scenarios range from 2–6 hours.

TORGAU, GDW (SPI 7·1, 7) Another game based on miniatures design, again admirable for its detail but slower than less precise games. Step reduction, with a clever system marking units in danger of routing with a different colour (no, not yellow). Movement can be in line or column, with changeovers time-consuming and possibly dangerous if enemy guns are nearby. Command control problems plague Frederick as he marches onto the battlefield, with a difficult dilemma between a safe approach and a fast one. 480 units, making a large but not unmanageable game.

TRIPLANETARY, GDW (SPI 6·5, 65) Back to space wars (*see* the games starting *Star* ... for alternatives), but with a relatively realistic scope; space ships manœuvre in the inner Solar System, with movement decided by vector forces, modified by planetary gravities; the system is not difficult, though it takes a little getting used to. 72 ship counters range from transports to battleships, and the scenarios start with a race game to demonstrate the movement system, working up to a struggle between rival planets, which gives a long and intricate game.

TSUSHIMA, GDW (not in polls) Brother of *Port Arthur*, giving the naval angle of the Russo-Japanese War of 1904–5; the Russians were confident that their European fleet, which sailed round the world to reach the spot, would carry the day, and that their flagship was unsinkable – both beliefs turned out to be false. Counters are at individual ship level except for destroyers and MTBs, and there are two boards: a map for movement at 100 miles to the hex, with zonal extensions reaching all the way to St Petersburg, and a combat board for when groups meet in action. Fire is simultaneous, and there are rules for running aground, mines, and breaking off engagements. *Tsushima* can be played with Port Arthur, or with abstract rules reflecting the course of the land action; the former seems to be the most interesting version, with the latter rather simple.

TURNING POINT, SPI (SPI 6·3, 93) SPI's game on the battle for Stalingrad; although this was probably the most crucial battle of the war, it is rarely simulated (AH's *Stalingrad* is really a game of the whole war on that front). The level is grand tactical, with 16-kilometre hexes and 180-plus counters, of divisional/corps size, with some air units. Game system resembles *Kursk* and *France 1940*, with a double movement phase for motorized units. Sixteen scenarios, with an optional rule executing Hitler's counter-productive order that no units should retreat. 3–4 hours, fairly complex.

UFO, JD (SPI 5·2, 176)

USN, SPI (SPI 6·7, 44) Strategic struggle for the Pacific, 1941–3, with 400 counters featuring land units at division/regiment level, air forces in multiples of ten planes, and naval groups, with aircraft carriers singled out individually. A long game, taking at least ten hours as a rule, although there are mini-scenarios; absorbing and highly complex. GDW have *Pearl Harbor*, on the whole 1939–45 Pacific War, coming out in the first half of 1977, and AH expect to bring out their version provisionally titled *The Rising Sun*, during the year; the former is by the designer of *Third Reich*, and the latter will have every capital ship singled out, so both games should join *USN* in the marathon class.

VERA CRUZ, SPI (not in polls) Operational level; invasion of Mexico, 1847. Out in September 1977.

VERDUN, Conflict (SPI 6·0, 119)

VICKSBURG, Rand (SPI 6·1, 109) Map runs from Commerce, Illinois, to Baton Rouge, Louisiana. Special importance of river flotillas, fortresses and supply depots is brought out. Union may build outflanking canals. Moderate complexity. Four scenarios and two campaign games. (cv)

VICTORY AT SEA, DCC–AWA (not in polls) Due out in 1977: air/sea tactical game from World War I to the present, with range determination and salvo patterns eliminating the usual die-roll.

VIKING, SPI (SPI 6·6, 54) *See Chariot*. The Crusades, the battle of Hastings, and struggles with the Mongols appear as well as Viking raids. Special rules deal with the Francisca weapon, Viking ferocity, and (in detail) Viking ships.

VON MANSTEIN'S BATTLES, Rand (SPI 6·5, 65) Rand's most popular game in the poll, with eight operational-level battles in the Ukraine, armour playing a major role. Exciting and highly playable.

WACHT AM RHEIN, SPI (not in polls) Highly detailed *battalion*-level simulation of the Battle of the Bulge, with the possibility of breaking down into companies! Untried units, divisional and non-divisional artillery, air power, engineers, paratroops, fortifications, anti-tank units, extensive supply rules; a real feast for the expert, ghastly for beginners. 44″ × 68″ map.

WAGRAM, SPI (SPI 7·0, 15) Part of the successful *Napoleon at War* Quad; fought on a large, open plain between two fairly equal armies. (cv)

WAR AT SEA, AH (not in polls) Formerly produced by Jedko's John Edwards, like *Russian Campaign*, with a semi-abstract treatment of the Second World War afloat which is very simple indeed, so the company suggests it can be played with wives, a singularly sexist conclusion! They do also suggest newcomers, kid brothers, and hard-core players in need of a break. There are forty-seven British capital ships, which the British player distributes around five areas; the Axis responds by committing U-boats and ships to the weaker Allied areas. There is no CRT, and just four pages of rules: the result is a pleasant little contest without being a fully-fledged wargame in the more sophisticated sense. Playing time 1 hour.

WAR BETWEEN THE STATES, SPI (not in polls) Sizeable brigade/division level simulation of the five-year campaign, with weekly (yes, weekly) turns, and a 66″ × 34″ map of the Confederacy and border regions. Weather, supply and leadership constraints prevent either side dashing for a quick victory. Some concepts reappear from *War in the West*, notably a monthly strategic cycle for strategic matters (recruitment, production, strategic supply, attrition, politics, foreign policy, the blockade). Most manœuvres done by corps or army formations, helping playability. *Cf. American Civil War* for a simpler game on the same subject.

WAR IN THE EAST, SPI (SPI 7·1, 7) The mighty rival to *Drang Nach Osten* plus *Unentschieden*, 208 turns and a quadruple-sized map. There are two editions, of which the more recent is preferable as it mates with *War in the West* and clarifies some rules. It is extremely hard to decide between this game and GDW's: both have every possible detail about the entire Soviet front at operational level, and real enthusiasts are unlikely to rest until they have tried both. However, it does seem to be a little more common that *War in the East* is actually played in practice; conversely, *Drang Nach Osten* does a little better in the poll, and is geared to the still more colossal Europa project. The extent and future scope of *Drang Nach Osten* is thus even greater than *War in the East*, but the SPI game is, I think, a little more playable. Really it's a matter of taste; the GDW game is somewhat cheaper if you leave out *Unentschieden*, otherwise not. Shorter scenarios, it should be noted, are available, each of around twenty turns.

WAR IN EUROPE, SPI (not in polls) The full *War in the East* plus *War in the West* package, with the rules to play them together as one grand game. Everything said about *War in the East* applies here as well, only more so. Shorter scenarios included. The air war is notably more abstract than in *Drang Nach Osten*, and individual land units are not named; play is smooth and practically free of paperwork.

WAR IN EUROPE, MODULE I, SPI (not in polls) An expansion kit of *War in Europe*, with new rules, charts and counters, enabling you to use the maps for the First World War instead – this time at corps level, however.

WAR IN THE PACIFIC, SPI (not in polls) Out in June 1977: the 1941–5 campaign, with 1600 air, sea and land counters, seven 23″ × 35″ map sections, and several scenarios as well as the full campaign game.

WAR IN THE WEST, SPI (not in polls) Actually, it was included in the following *Strategy and Tactics* poll (issue 58), and unsurprisingly went straight to the top, with a thumping 7·6. Again the same comments apply, except that in this case there is at present no rival game covering the same extent in the same depth. Various shorter scenarios, including a four-turn game on Poland in 1939, give players a good taste of the full design.

WAR OF THE WORLDS II, Rand (SPI 4·3, 198) Ill-received science-fiction game on interplanetary conflict in the year 2000. Two to three players jostle to establish and control colonies on five planets.

WARSAW PACT, Jagdpanther (not in polls) Corps/divisional level simulation of possible wars in Europe in the next twenty years. 25 miles per hex; 5 days per turn; 10-turn game. Combat pretty bloody; reduced strength on back of unit counters. Russians limited with supplies which may be used to boost attack strength, etc. Airmobile, airborne and marine operations are included. All Nato and Warsaw Pact nations are represented, with neutrals such as Switzerland and Yugoslavia. Scenarios run the gamut from limited wars in the Balkans to The Big One. Frighteningly realistic. (cv)

WAR OF THE STARSLAVERS, DCC–AWA (not in polls) Scheduled for

1977 release, this is a multi-player game featuring slavers, pirates and two warring empires. Spaceships cost money to maintain, and the object of the game is to emerge with a profit, however fiendishly you have to stab allies to do it.

WATERLOO, AH (SPI 5·8, 136/AH 5·82, 19) Another in the classic line, with plenty of excitement and action as usual; a recent new edition of the rules eliminated some old oddities. Second only to *Tactics II* and *Origins of World War II* in the separate AH poll on 'ease of understanding', but weak on realism, especially with the absence of a special role for artillery. The *Napoleon's Last Battles* Quad, taken together, make a good, not much more complex, alternative. Speaking of which ...

WAVRE, SPI (not in polls) Part of the afore-mentioned Quad. Grouchy versus Blücher, each trying to delay the other from reinforcing the main battle. The varied map is fully used.

WELLINGTON IN THE PENNINES, Rand (SPI 6·4, 87) Area movement game of the Peninsula War. Units come in 'strength points', reducing period glamour. Leaders, siege trains, and occupation forces appear. The CRT works well, but attrition by guerillas is rather tedious. Nine scenarios. Smooth game system but lacking in flavour and fun. (cv)

WELLINGTON'S VICTORY, SPI (not in polls) Battalion-level simulation of Waterloo, with 1600 counters, and a mere hundred yards per hex. Miniature-influenced design with columns, squares, extended lines and skirmishers. Movement mostly in brigade-sized columns, fortunately for playability, but it's still an imposing game.

WEST WALL, SPI (SPI 7·0, 15) Tops all the Quads in the poll, with *Arnhem*, *Hurtgen Forest*, *Bastogne*, and *Remagen* featuring; it is difficult to design good, simple games of modern combat, because modern combat is very far from simple, but *West Wall* brings it off, with a modification of the *Modern Battles* system. *Arnhem* is especially impressive.

WHITE BEAR AND RED MOON, Tactical Studies Rules (not in polls). Fantasy board wargame by the makers of the celebrated miniatures game *Dungeons and Dragons*. Eight scenarios of increasing complexity; the total effect is highly complicated. Bloodthirsty CRT with a good deal of luck, but absorbing diplomatic possibilities if played with a number of participants.

WILDERNESS CAMPAIGN, SPI (SPI 5·5, 160) Lee and Grant meet in 1864; the game divides into two scenarios, one up to and including Cold Harbor, and the other afterwards. The design is based on *Franco-Prussian War*, with new emphasis on cavalry reconnaissance, river and sea combat and effective leadership. The game is balanced with the help of 'what-if' options. See *Battle of the Wilderness* for a tactical treatment; this game has a more strategic approach. 3–4 hours, fairly complex.

WINTER WAR, SPI (SPI 6·4, 80) Exciting if rather luck-dependent game of the Soviet war with Finland shortly before the Second World War. The main action is in assaults on the fortified Finnish lines in the south, but Finnish ski patrols skirmish further north with Russian units trying to turn the defending flank. Tends to be biased to Finns, unless the optional rules are used without the cease-fire option. 2–4 hours and good fun. (*See* Chapter 3.)

WOLFPACK, SPI (SPI 5·4, 169) Solitaire-only game, with U-boats hunting convoys and dodging escorts moving in unpredictable patterns. *Operation Olympic* is generally thought a better solitaire game, unless you have a strong preference for submarine warfare.

WOODEN SHIPS AND IRON MEN, AH (SPI 7·1, 7/AH 6·66, 1) Avalon Hill's top-scorer in both polls, a development of a Battleline design. Playable yet highly realistic, it simulates numerous naval battles at a detailed tactical level, with different types of sail, various forms of ammunition, and an immense range of different ships; wind plays a crucial role, and the reasons for the historical formations become apparent. Grappling and boarding are common, with victory depending on morale and the number of survivors from the previous rounds of fire. Many scenarios are unbalanced, but experienced players design their own groups from a points system. Complex but accurate and clear rules. Scenarios from thirty minutes to many hours.

WORLD WAR I, SPI (SPI 6·2, 102) One might expect this to be one of the 1000-unit, 100-rule variety, but it isn't: a small map squeezes in every front in continental Europe, and extremely ingenious rules encompass the whole war at a grand strategic level, enabling the game to be played in a few hours. Casualties are exacted from national manpower reserves, from which new units are also also drawn. The Central Powers have a very difficult time (which imbalance may account for the relatively low rating), as the Allies get a big bonus for their naval power automatically. *Stosstruppen* stiffen the Germans in the last years, but too late unless Russia has been decisively knocked out of the war earlier. Exciting and highly playable; moderately complex; realism a bit doubtful at this eagle's-eye-view level.

WORLD WAR II, SPI (SPI 6·6, 54) The rival to *Third Reich*, dealing with the whole war, though only on the European fronts. Seasonal turns and complex rules, though the total effect is a little simpler than the AH game, with a correspondingly somewhat narrower scope – thus, the production system which is central to *Third Reich* is absent here.

WORLD WAR III, SPI (SPI 6·4, 80) Again, not the giant one might expect, but a playable simulation of a nasty possible future. The Russians are billed as the culprits, and are able to steam-roller Western Europe with a gradually diminishing advantage from surprise which nevertheless lasts a full year. Japan can probably hold out, together with a Norwegian foothold (very implausibly). The backbone of the game is the innovative naval aspect: a vast fleet of Western merchant shipping keeps Western hopes alive while the East Bloc navies are herded off the oceans, and the US can gradually prepare a massive armada to return to Europe. The game suffers from very simple victory conditions, which ensure that victory hinges entirely on control of a few industrial centres, while whole continents are more or less untouched. I recommend planting some industry in South Africa, Australia and Brazil, and loosening the supply rules to make Russian invasions overseas easier. This greatly enlivens the game, but maintaining play-balance is tricky. 5–8 hours, with two rather lacklustre minigames (sea-only and land-only). Nuclear rules are provided but the designer sensibly advises against their use.

WURZBURG, SPI (SPI 6·9, 23) This game had a spectacular launching in West Germany: on hearing that it simulated a Soviet invasion of the country in which the Allies used nuclear weapons around the town, local politicians erupted in fury, and a prominent German weekly wrote a scathing article about people who have fun blowing up Wurzburg. The German distributors recoiled, appalled, and took the game and other modern-era games off the market. Arguably, it would be more productive to worry about nuclear weapons than about games which refer to them. Whatever view one takes on this, *Wurzburg* is a good game, part of the *Modern Battles* Quad, with a fierce series of typical battles around the town. (cv-np)

YEAR OF THE RAT, SPI (SPI 5·5, 159) Divisional-level simulation of the North Vietnamese/NLF 'Tet' offensive which transformed public opinion on the state of the war, though failing in purely military terms. North Vietnamese units start inverted along the borders of the south, so that the Saigon player is never sure where the main thrust is coming, and they have a strong advantage off the main roads. However, the Saigon ARVN forces have increasing American air support, and are able to impose severe supply problems on the attackers. Fairly well-balanced and skill-demanding on both sides (especially the Communists, who lose heavily in all straightforward fights in the open), but not wildly exciting. 2–3 hours, moderately complex.

YEOMAN, SPI (SPI 6·7, 44) *See Chariot*. Bannockburn, Crecy, Agincourt, and Biococca feature in the pre-gunpowder Renaissance period. Squares, foolhardy cavalry, longbowmen, trenches, cavalry traps, and artillery limbering appear, giving fair period 'feel'.

THE YTHRI, Metagaming Concepts (SPI 6·0, 119) Lively, simple simulation of the war described in Paul Anderson's book, *The People of the Wind*; mainly suitable for readers of this.

YUGOSLAVIA, SPI (not in polls) Part of the *Modern Battles II* Quad out in mid-1977, featuring US intervention against a Soviet attack.

BEST 39 GAMES IN SPI'S POLL (OF 202):

1) *Drang Nach Osten*. 2) *Bataille de la Moskowa*. 3–11) *War in the East, Frigate, Antietam, Wooden Ships and Iron Men, Dreadnought, Barlev, Jena-Auerstadt, Torgau, Crimea*. 12–19) *Sinai, Shiloh, Chickamauga, Wagram, Global War, Kingmaker, Chinese Farm, West Wall Quad*. 20–26) *Blue and Grey Quad (I), Wurzburg, Borodino, Napoleon at War Quad, Battle of Nations, Narvik, Spartan*. 27–39) *Panzerblitz, Panzerleader, Modern Battles Quad, Sniper, Starforce, Third Reich, Diplomacy, Patrol, Cemetery Hill, Golan, Musket and Pike, Panzer 44, MechWar '77*.

BEST 5 GAMES IN AH'S POLL (OF 25):

1) *Wooden Ships and Iron Men*. 2) *Anzio*. 3) *Panzerleader*. 4) *Richthofen's War*. 5) *1776*.

RECOMMENDED BY THE AUTHOR AS INTRODUCTORY GAMES

From SPI: *Napoleon at Waterloo* (basic and expanded), *Borodino*, all Quad games.
From AH: Any classic game, especially *Waterloo* and *Afrika Korps*, *Luftwaffe* (Basic game).
From other companies: *Chaco*, *Cromwell.*

CANDIDATES FOR THE ULTIMATE SIMULATION

From SPI: *War in the East*, *War in the West*, *Wellington's Victory*, *War in the Pacific.*
From GDW: *Drang Nach Osten/Unentschieden*, *Avalanche*, *Pearl Harbor.*
From AH: Forthcoming Civil War and Pacific games.
From other companies: *La Bataille de la Moskowa*, *Shenandoah.*

PERSONAL FAVOURITES OF THE AUTHOR

Avalanche, *Battle of the Bulge*, *Cromwell*, *D-Day*, *Dreadnought*, *Luftwaffe*, *Seelöwe*, *Sinai*, *Sixth Fleet*, *Thirty Years War Quad.*

SOLITAIRE GAMES:

From SPI: *Operation Olympic*, *Wolfpack*, *Fall of Rome.*
Numerous 2-player games can be played solitaire.

PART V

Sample Game

THE BATTLE OF NORDLINGEN 6 SEPTEMBER, 1634

Prologue: We will conclude the book with an example of a full game played out turn by turn: *Nordlingen*. First you should re-read the description of this game and its rules in Chapter 7; the starting positions are also shown there. Illustration 23 shows the combat results table and artillery fire table. Note the following rules in particular:

1) Disrupted units are eliminated if disrupted a second time. Leaders (with asterisks on the counter) add their value to the die-roll when disrupted units try to rally, which succeeds on a result of 'five' or 'six' – thus a leader with value 1 will enable disrupted units to rally (undisrupt) on a roll of 'four', 'five', or 'six', and one of value 2 will undisrupt units rolling 'three' as well. Leader effects work within one hex of the leader.
2) Artillery fire cannot go *through* forest, town, or hilltop hexes, or hexes occupied by anything more than isolated leaders. Thus, the Imperialist rear line is protected from the Swedish guns by its front line. Artillery fire does not affect leaders or artillery. Guns on hilltops (the Imperialist guns) can ignore blocking terrain if it is nearer them than the target. Artillery fire and disruption removal are at the start of each turn, followed by movement and combat.
3) Zones of control force combat and prevent enemy disruption removal, but have no effect on movement.
4) Only artillery (which cannot move) and leaders can stack with other units. Leaders cannot fight on their own.
5) Defenders attacked up slopes are doubled. Swedish infantry and leaders are tripled on attack on the first turn; this only helps Horn's

[12.0] TERRAIN EFFECTS CHART

(See Terrain Key on Map.)

Terrain	Movement Points [MP] to Enter [or Cross]	Effect on Combat
Clear Hex	1 MP	No effect.
Forest Hex	May not enter.	Not allowed.
Woods Hex (Nordlingen only)	2 MP. Cavalry may not enter.	No effect. Cavalry may not attack units in Woods Hex.
Town Hex	May not enter.	Not allowed.
Road Hex	1 MP; negates effects of other terrain in hex if hex is entered through Road Hexside.	No effect.
Slope Hexside	1 MP additional moving from Slope Hex; no additional MP moving into Slope Hex.	Defender doubled if all attacking units attack across Slope Hexside from Slope Hex.
Stream Hexside	2 MP additional to cross.	Defender doubled if all attacking units attack across Stream and/or River Hexside.
River Hexside (Nordlingen, Freiburg only)	May cross at bridges only.	May only attack across bridges.
Bridge Hexside	No additional MP	Defender doubled if all attacking units attack across bridge (or ford) hexsides.
Marsh Hex (Breitenfeld only)	2 MP	Combat Strength of units in hex halved (fractions rounded up).
Rough Hex (Nordlingen only)	3 MP. Cavalry may not enter.	No effect. Cavalry may not attack units in Rough Hex.
Ditch Hex	5 MP for Cavalry; 2 MP for non-Cavalry.	No effect.
Entrenchment Hex (Freiburg only)	Cavalry may not enter. 1 MP for Bavarian non-Cavalry; 3 MP for French.	Defending Strength of Bavarian units increased by 1 Strength Point. Cavalry may not attack.
Fortification Hexside (Freiburg only)	May not cross unless breached; then pay 1 MP additional to cross.	Not allowed unless breached (see 19.2); then defender doubled if attacked.
Gate Hexside (Freiburg only)	May not cross unless Friendly; then no additional MP to cross (19.27).	Not allowed unless breached (see 19.2); then defender doubled if attacked.

[13.0] DESIGNER'S NOTES

(See Exclusive Rules Folder.)

[5.1] ARTILLERY FIRE TABLE

Die Roll	Range in Hexes: Artillery counter to Target 1	2	3-5	6+	Die Roll
1	Dd	Dd	Dd	Dd	1
2	Dd	Dd	Dd	•	2
3	Dd	Dd	•	•	3
4	Dd	•	•	•	4
5	•	•	•	•	5
6	•	•	•	•	6

[5.11] Explanation of Artillery Fire Table

The Artillery Fire Table is divided into four columns corresponding to range between the firing Artillery counter and the target unit's hex. For the purposes of the game, the range of the Artillery is unlimited; however, the effectiveness of Artillery fire does vary inversely with the range. To determine the range, count the number of hexes between the Artillery counter (exclusive) and the target unit's hex (inclusive). Then cross-reference the die roll with the range to find the result. The two results possible on the Artillery Fire Table are "Dd" and "•." "Dd" = Disruption (see Section 9.0); "•" = no effect. NOTE: A unit may **never** be eliminated as a result of Artillery fire (i.e., Artillery fire has no effect upon disrupted units).

[8.6] COMBAT RESULTS TABLE

Die Roll	Probabilty Ratio (Odds) Attacker's Strength to Defender's Strength 1-5	1-4	1-3	1-2	1-1	2-1	3-1	4-1	5-1	6-1	Die Roll
1	Ad	•	•	Dx	Dd	Dd	Dd	De	De	De	1
2	Ad	Ad	•	•	Dx	Dd	Dd	Dd	De	De	2
3	Ae	Ad	Ad	•	•	Dx	Dd	Dd	Dd	De	3
4	Ae	Ad	Ad	Dx	•	•	Dx	Dd	Dd	Dd	4
5	Ae	Ae	Ad	Ad	Dx	•	•	Dx	Dd	Dd	5
6	Ae	Ae	Ae	Ad	Ad	Dx	•	•	Dx	Dd	6

Attacks executed at Odds greater than "6-1" are treated as "6-1;" attacks at Odds lower than "1-5" are treated as "1-5."

[8.61] Explanation of Combat Results

Ad = **Attacker Disrupted.** All attacking units are disrupted (see Section 9.0).

Dd = **Defender Disrupted.** All defending units are disrupted. Defending units which were previously disrupted and against which this Combat Result is achieved are eliminated.

Dx = **Disruption Exchange.** All previously undisrupted defending units are disrupted. All previously disrupted defending units are eliminated. The attacking Player must disrupt attacking units whose total printed Combat Strength is **at least** equal to the total printed Strengths of all the Defending units. Only units which participated in the particular attack in question may be so disrupted.

De = **Defender Eliminated.** All defending units are eliminated (removed from the map).

Ae = **Attacker Eliminated.** All attacking units are eliminated.

• = **No effect.**

23 TEC, CRT and Artillery Fire Table from *The Battle of Nordlingen.*

southern group assaulting the Allbuch, as the main forces cannot meet on the first Swedish turn, being 9 hexes apart.

6) When *total* losses on their side reach the level shown, the groups referred to become demoralized and when disrupted are eliminated except where stacked with leaders – a shattering blow as there are only three leaders on each side. If one side has a category demoralized, the other increases its demoralization levels by 25 points. Infantry losses count towards cavalry demoralization, and vice versa.

Imperialist Cavalry: 100
Imperialist Infantry: 125
Swedish cavalry: 75
Swedish infantry: 100

7) Charging Swedish cavalry are doubled against disrupted infantry, but are themselves disrupted as a result.

8) Victory points are gained thus: ten per artillery piece captured, five per leader value point eliminated, two per infantry point killed, and one per cavalry point killed. Because of the demoralization rules, it will usually be quite clear who is winning.

The Swedes move first. The cost of two movement points instead of one for crossing a slope uphill prevents them gaining the heights of Allbuch on Turn 1, also the defence is doubled. The urgent requirement is to disrupt the enemy; Swedish disruptions at this stage can be quickly rallied, while the Imperialists are unable to get out of Swedish ZOCs without abandoning the artillery, so their disruptions are likely to stay. Because of the stacking limit of one, only three of the four infantry units can reach the front; the other 13-4 slips round the side for use next turn. The 5-8 cavalry joins Horn (the 2-8 leader) and the southernmost 15-4 in one of the attacks on a doubled 5-3, although this only increases the odds from 15 plus 2 tripled = 51–10 to 56–10, still 5–1, the reason being that a 'disruption exchange' (see the CRT) is possible, and then the Swedes will prefer to disrupt their 5-8 rather than the 15-4. The other 15-4 and 13-4 attack at the maximum odds, while the flanking cavalry gets attacks at 1–1 and 1–2.

The result is that one of the four enemy units is eliminated, two more are disrupted, and the fourth is unaffected; the Swedish 7-8 is the only attacker disrupted (in a disruption exchange). Quite a satisfactory attack – so far, so good!

The main force sends out probes in both directions, after firing artillery at long range with a single disruption achieved on a 5-3 on a hilltop at the southern end of the main enemy line (useful since this might have reinforced the Allbuch). A powerful group, including both leaders in the main force, mass northwest of Klein Erdlingen, ready to fight

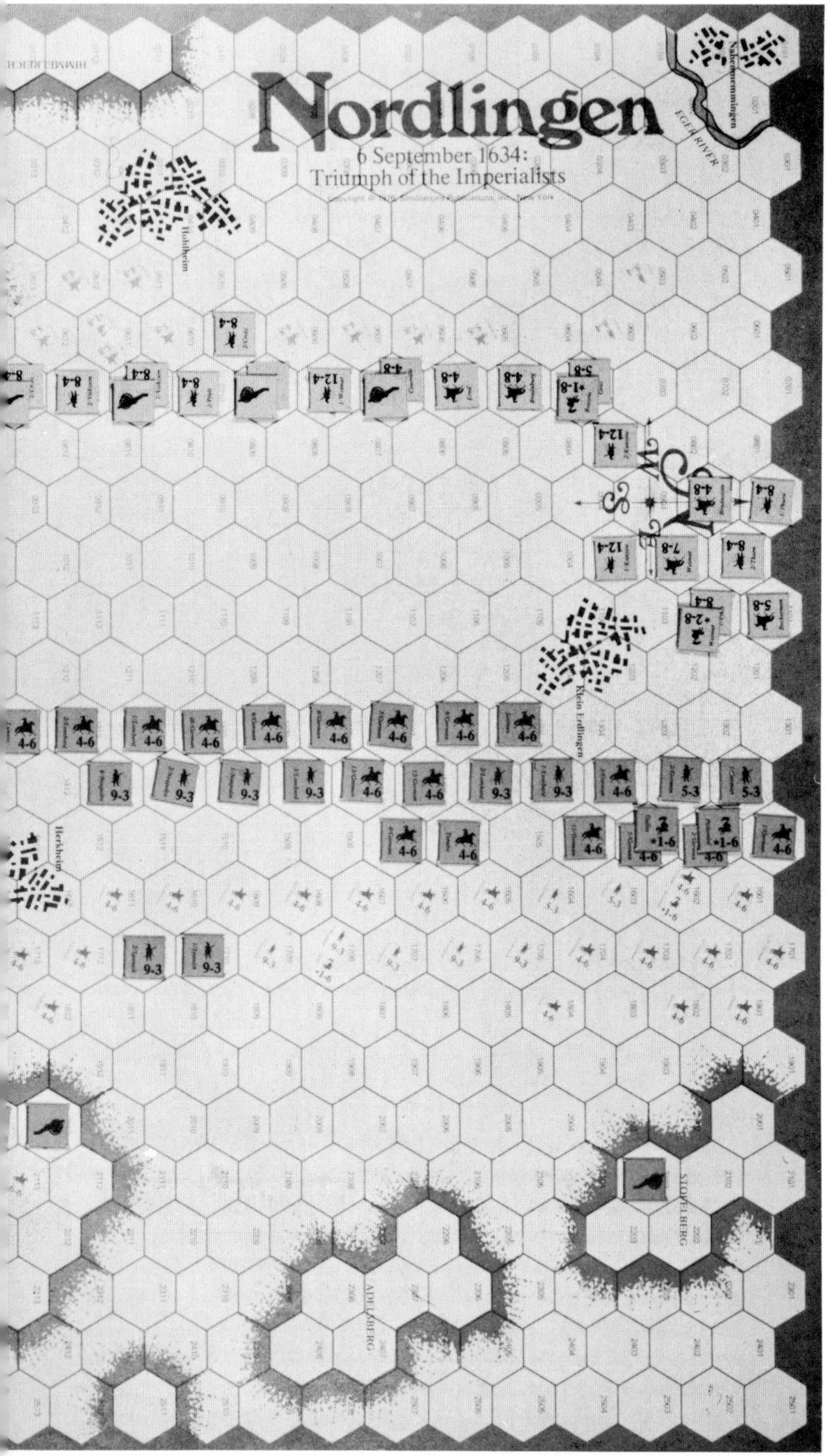

Nordlingen
6 September 1634:
Triumph of the Imperialists
HIMMELREICH
Hohlheim
Nahelemmingen
EGER RIVER
Klein Erdlingen
Herkheim
STOFFELBERG
ADELBERG

24 Position after turn 1.

the enemy cavalry if they push forward here. Meanwhile, four cavalry units set off south, avoiding the forest and rough terrain, which is barred to cavalry (and slow for infantry). These fast units can ride back to the main front if the enemy sends massive reinforcements south, and carry on to the Allbuch if he does not; the Swedish mobility is fully used here to gain an advantage. Note that the Swedes have effectively surrendered the option of a main force clash in the centre by these diversions, but both players know that this is unlikely to pay for the Scandinavians anyway.

Illustration 24 shows the position after the first Imperialist turn (the Time Record marker is advanced to the next turn). He failed to undisrupt his 5-3 on the Schonfeld (the other disruptions cannot be removed, because of Swedish ZOCs). Disrupted units are shown inverted with their reduced combat strength in brackets. His main force artillery could not fire north of Klein Erdlingen, because his own intervening units were nearer than half way from the guns. The five guns therefore fire on the Swedish centre, disrupting a 12-4. In the south, the two Allbuch guns fire at point-blank range, and are a little fortunate in disrupting a 15-4 and a 13-4. The undisrupted Imperialist 5-3 moves up next to them to stop them rallying next turn, and one of the disrupted defenders moves onto the other gun to prevent its capture. The third (disrupted) defender surviving moves its maximum two hexes away, in the hope of being out of harm's way next turn. The undisrupted 5-3 is forced to attack the 5-8 at 1–1, and rolls an exchange (of disruption – 'exchange' in the sense of removing units does not exist in this game), so it is unable to prevent the enemy infantry staying undisrupted after all; by being disrupted itself it loses its zone of control. The disrupted units are inverted, with their reduced defence values shown on the back.

A large Imperialist force moves south to reinforce the Allbuch, and a wedge of units moves southwest to cut off the Swedish cavalry reinforcements if they press on; if the Swedes return to the main front, the wedge can switch to it as well without losing too much time. In the centre, the cavalry has not advanced nearer than six hexes from the Swedish line, because it is going to wait for the infantry, and going closer would increase the Swedish artillery effectiveness (see the artillery fire table, illustration 23). In the north, no attempt was made to assault the powerful Swedes on the narrow front behind the town (into which neither side can enter), but infantry is within striking range for next time.

Losses after turn 1: Imperialists 5 factors, Swedes 0.

On the second Swedish turn, the Swedes manage to undisrupt only two units, and the guns blaze away to no effect. Horn's forces surround the surviving Imperialist defence and destroy them at maximum odds (with forces in their rear, they are no longer doubled for hill defence). The disrupted units are placed where they will still be out of enemy ZOC next turn. But the guns will not be captured until next turn, so the Imperialist sacrifice will not be in vain.

In the centre and north, the Swedes decline combat; the only place where it is worth considering is the north, and here the infantry would be needed to make an impact, and would be quickly cut off from being able to reinforce the centre when needed. The cavalry moving south presses on cautiously.

Imperialist turn 2: the situation after this is shown in illustration 25. Only part of the main artillery can still fire at anything after the advance of their troops; these do so to no effect. The Allbuch guns fire their last Imperialist barrage, disrupting one of the 15-4s but not the other. A leader helps one disrupted Imperialist 5-3 to regroup, but the one which escaped from the Allbuch slaughter is still (understandably) too shaken to rally.

The fresh units reach the Allbuch and attack the disrupted 15-4, with a soak-off on a (doubled) 4-8. The soak-off gives an 'attacker disrupted' result, but the main battle results in an exchange, disrupting the attackers but eliminating the already disrupted Swedish unit – a major success. Another disruption exchange occurs southwest of Klein Erdlingen, while the remaining clashes give no effect. The main central forces press on, near the Swedish lines. Again, the cavalry screen avoids advancing closer than necessary.

Total losses after turn 2: Imperialists 15 factors, Swedes 15.

Swedish turn 3: the leaders help two disrupted units rally, but a 13-4 by Horn stays disrupted. The Swedish artillery, now at range 3, disrupts two enemy cavalry units. A counter-attack starts at the flank of the guns, southwest of Klein Erdlingen, with a view to knocking out two disrupted units; the first Swedish cavalry charge occurs here, destroying the 9-3 which obtained a disruption exchange on the Imperialist turn. One of the disrupted cavalry is also destroyed, but in an exchange which disrupts a Swedish 8-4, not a good bargain. A 3–1 attack on another cavalry piece by one of the Swedish 12-4s has no effect.

In the south, the fresh cavalry arrives, and helps savage the advanced Imperialist cavalry, but a slightly risky attack by a 13-4 at 4–1 (which could have led to disruption – by exchange – and almost certain elimination on the enemy turn) simply has no effect. The Allbuch guns are captured and can now fire for the Swedes.

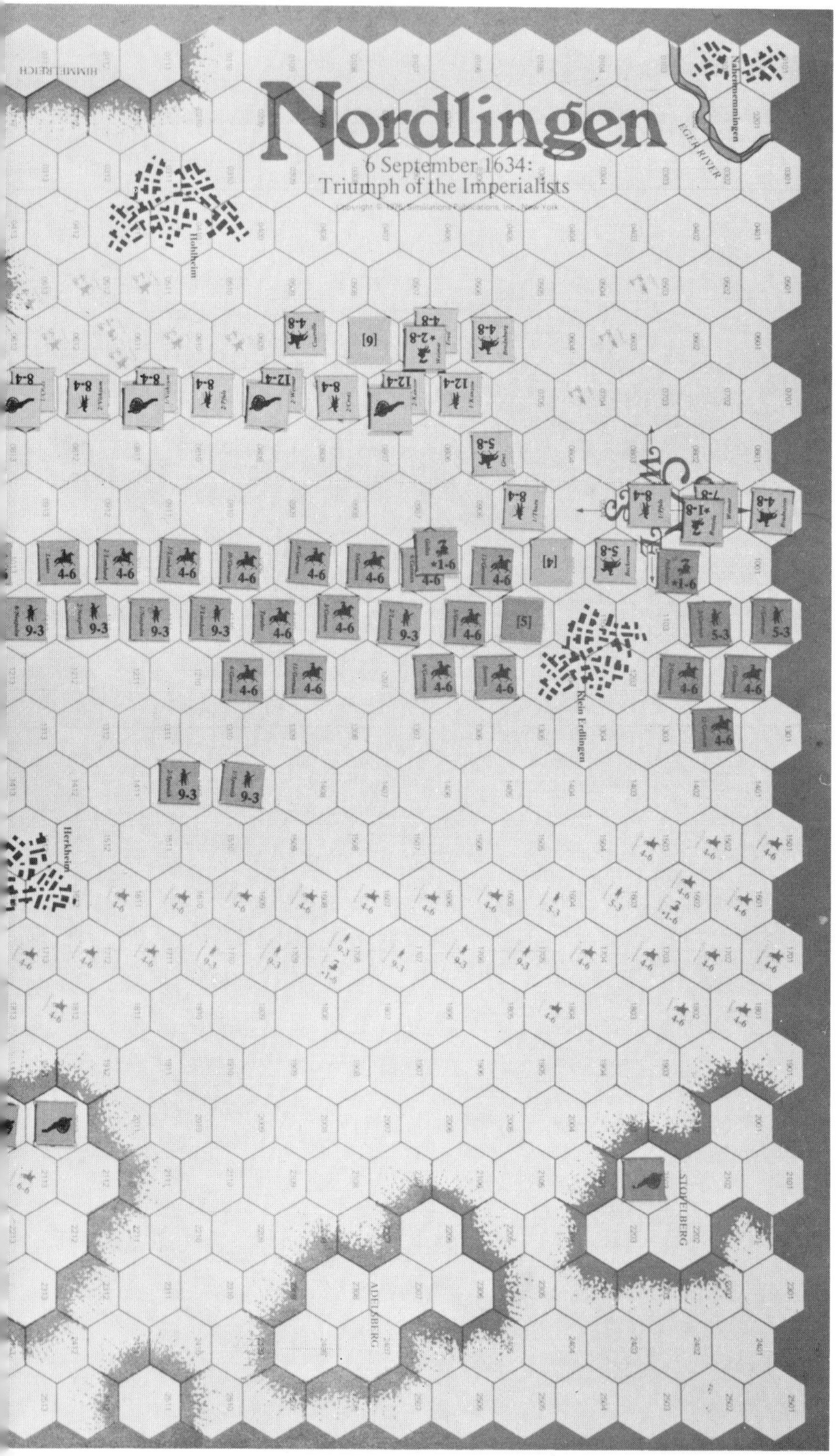
Nordlingen
6 September 1634:
Triumph of the Imperialists
HIMMELREICH
Hohlheim
Nahermemmingen
EGER RIVER
Klein Erdlingen
Herkheim
STOFFELBERG
ADELBERG

25 Position after turn 2. (The 1-6 Imperialist leader on hex 1002 is really stacked with the 5-3 on the hex behind it.)

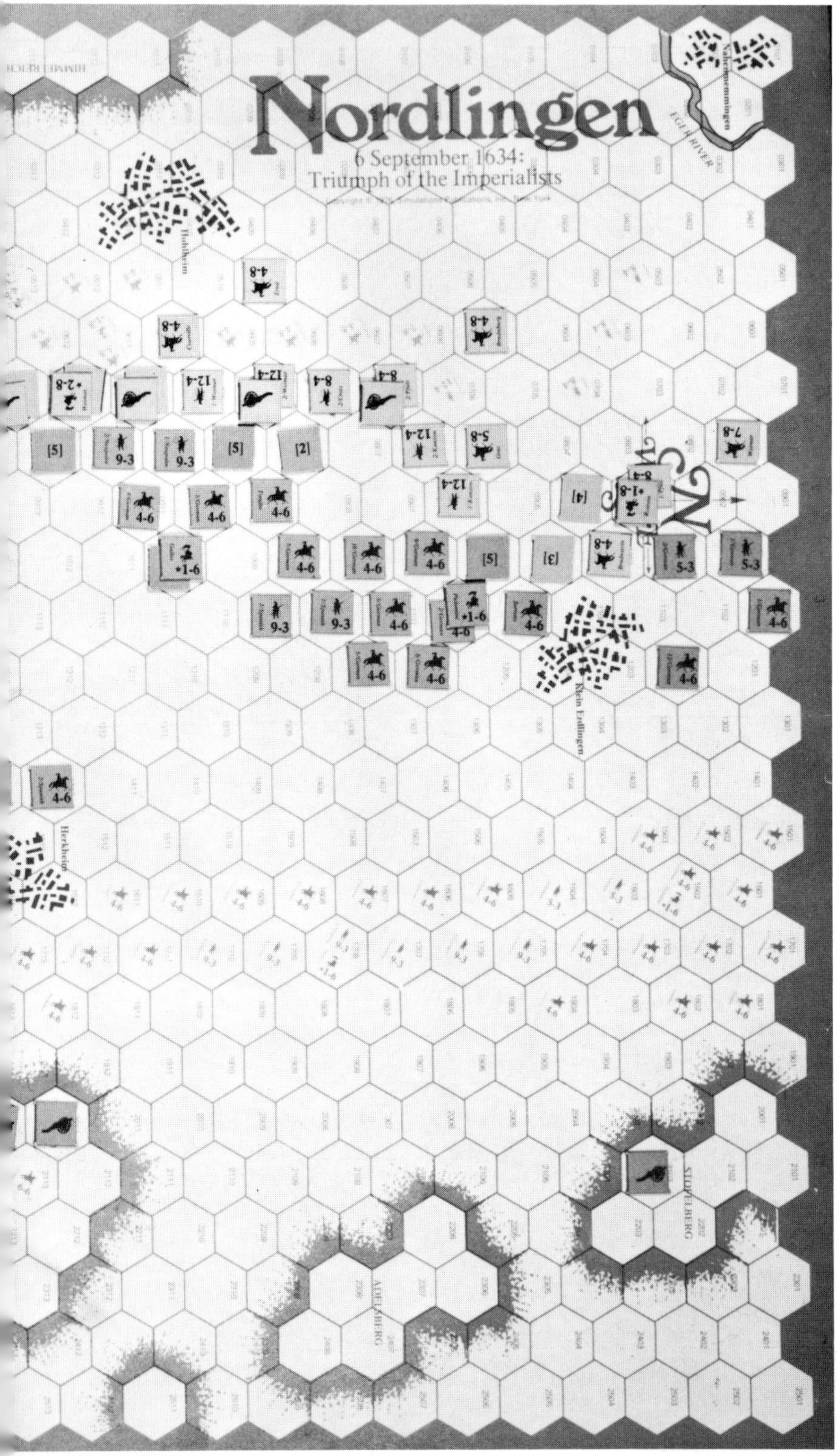
Nordlingen
6 September 1634:
Triumph of the Imperialists
Klein Erdlingen
Herkheim
Hohlheim
STOFELBERG
ADELSBERG

26 Position after turn 3.

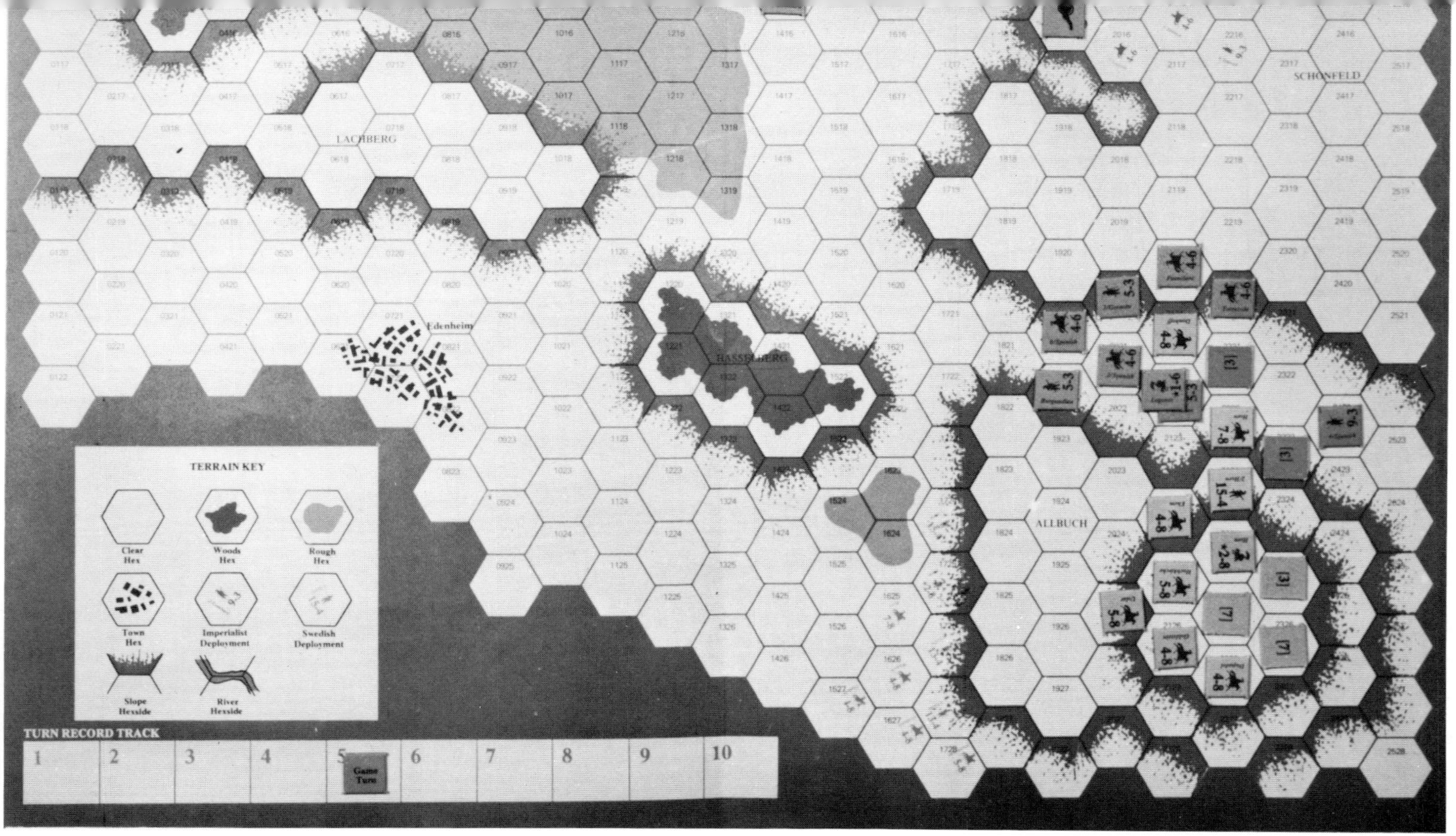

LACHBERG
SCHONFELD
Edenheim
HASSELBERG
ALLBUCH
TERRAIN KEY
Clear Hex
Woods Hex
Rough Hex
Town Hex
Imperialist Deployment
Swedish Deployment
Slope Hexside
River Hexside
TURN RECORD TRACK
1
2
3
4
5
6
7
8
9
10
Game Turn

27 Position after turn 4.

Imperialist turn 3: *see* illustration 26. One disrupted unit rallies. In the south, things are not going well, but the attack is pressed home as well as possible; four cavalry units from the centre move into a reserve position by Herkheim, ready to ride south or return to the central battle as the need arises. In the centre, the infantry storms in on the Swedish line, obtaining three disruptions for two disrupted attackers (plus one cavalry soak-off). Southwest of Klein Erdlingen, the 8-4 which the Swedes had disrupted on their turn is knocked out in an exchange, but the cavalry survives. In the south, a Swedish cavalry unit is unexpectedly eliminated, and several major forces are disrupted, with the Imperialists coming off worse. However, their 6-6 prevents the enemy 13-4 at the back of the Allbuch from undisrupting.

Total losses after turn 3: Imperialists 36, Swedes 27.

Swedish turn 4: although the Swedes have taken lighter losses so far, they have got their lower demoralization levels to worry them. However, they attack most of the disrupted units on the Allbuch, using a cavalry charge in one case. The 4-8 off the hill slips behind the enemy lines to prevent rallying, but does not attack, to avoid being disrupted, and then destroyed on the enemy turn. The southern artillery, captured last turn, could not fire due to blocking Swedish units. The Swedes regroup on the upper plateau behind the guns.

In the centre, the artillery disrupts another enemy 9-3, but otherwise fails; one gun is blocked by two Imperialist units which are already disrupted – artillery cannot affect these. The Swedish infantry assault the two undisrupted enemy units in the front line, and the cavalry charges two of the disrupted attackers in front of the guns. Southwest of Klein Erdlingen, the battle grows in fury, as the Swedes seize the chance to destroy the disrupted 9-4. Reluctantly, they advance in front of one gun, since they cannot hold the unbroken line otherwise; if a gap appears, they will be unable to have a safe area to undisrupt units. In the north, they decline combat; it might lead to disruption exchange and losses on the enemy turn, and the Swedes cannot afford to lose factors more quickly than they have to.

Imperialist losses are massive: another 36 factors! But many of the Swedish attacks (including, necessarily, all the cavalry charges) result in disruption.

Imperialist turn 4: the Imperialists tacitly concede the Allbuch; the units massed there behind the guns are impregnable. If the Swedes come out, that will be another matter. In the meantime, the southern forces concentrate on surrounding the loose 4-8 and eliminating a disrupted Swede. The Swedes stand to gain 20 points from the capture of the guns, so perhaps the Imperialists should have sent more south. However, in the centre, they are beginning to get the upper hand, and several disrupted Swedes are eliminated, with a number of fresh disruptions on both sides. Now that the Swedes have virtually no central reserves, the Imperialists feel they can fling in the assault all down the line without fear of general counter-attacks. *See* illustration 27 for the resulting position.

Total losses after turn 4: Imperialists 72, Swedes 60.

This makes turn 5 crucial for both sides. If the Swedes cannot eliminate 28 Imperialists, then it seems extremely likely that the Imperialists will eliminate 15 Swedes, demoralizing the Swedish cavalry and boosting their own morale by 25 points. This forces Horn's troops out of their hard-won stronghold, and they charge down the slope at their disrupted enemies. In the centre, the Swedish line disintegrates, though the guns are all guarded, in a desperate attempt to give the Imperialist morale the vital push over the brink. A preliminary round of artillery disrupts one more cavalry unit. In the ensuing fighting, 35 Imperialists are eliminated, and Imperialist cavalry morale breaks.

Illustration 28 shows the position after this Swedish turn. The Imperialist can try a last desperate offensive – the Swedish leader is vulnerably sitting on a disrupted unit – but his position is resignable. Even if, with extraordinary luck, he pushes the Swedish casualties over 100 this turn, demoralizing the Swedish cavalry, he will have so few units left that he will have no chance of capturing the guns; with the loss of two guns, and more casualties than his opponent, even killing Saxe-Weimar will not save him.

POST MORTEM

I chose *Nordlingen* for a demonstration because the decision tends to come quickly and decisively. The game, short though it was, illustrates most of the main concepts. The Swedish strategy of keeping the main body behind the guns was vindicated; the Imperialists *were* winning here, despite the tremendous defensive advantage of the Swedish artillery at close range, and if the Swedes had gone out to meet them in the middle, their losses would have been heavier and – even more important – earlier. In the game, because of the delay in the main battle, the southern front was responsible for many of the casualties, and here the Swedes had the edge.

The Imperialists' strategy was reasonably sound, though they guessed wrong on the amount of reinforcements needed in the south; in retrospect, they would have been better advised to send either more or fewer reinforcements south. But Horn's attack was a shade fortu-

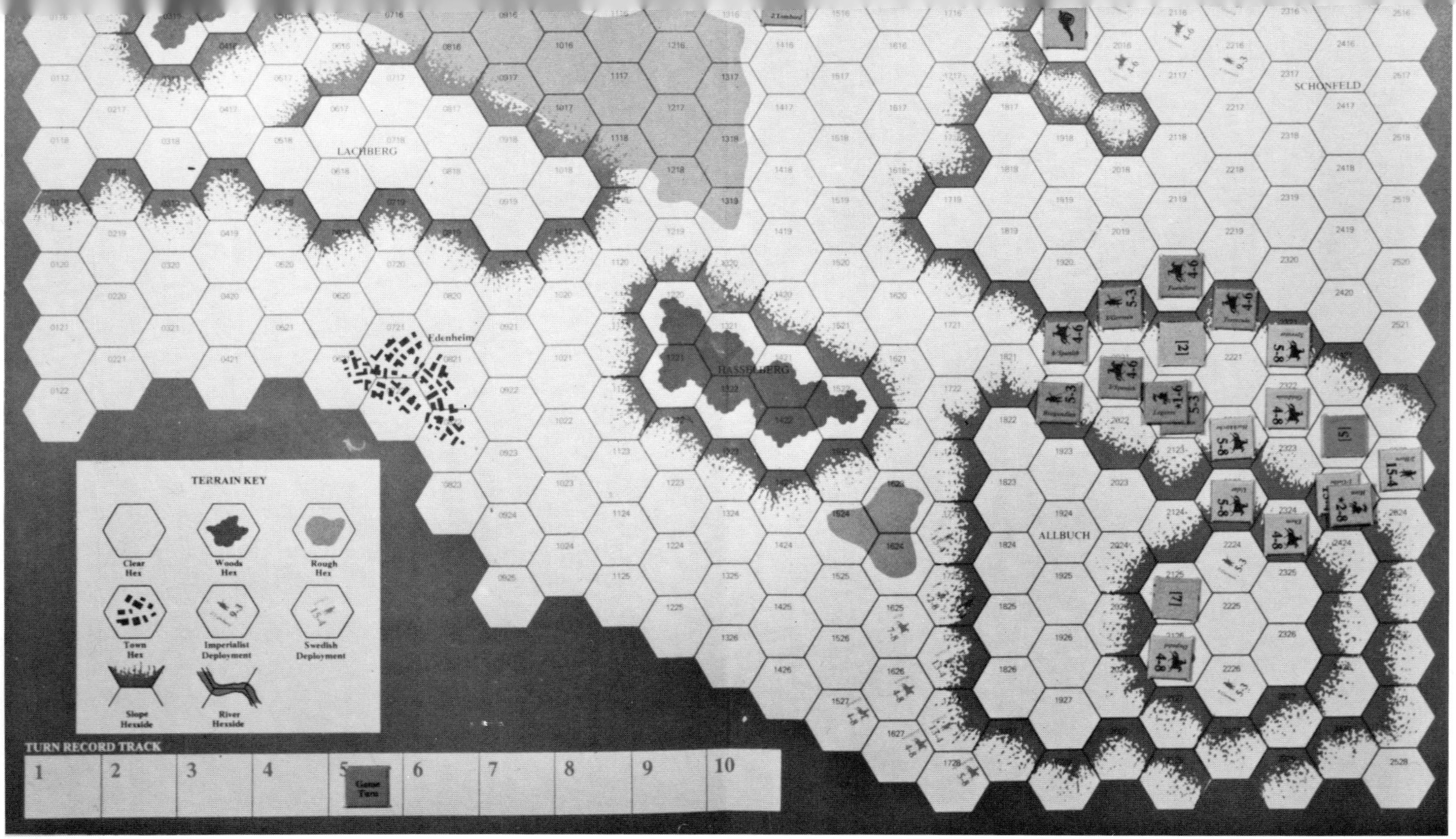
LACHBERG
Edenheim
HASSELBERG
ALLBUCH
SCHONFELD
TERRAIN KEY
Clear Hex
Woods Hex
Rough Hex
Town Hex
Imperialist Deployment
Swedish Deployment
Slope Hexside
River Hexside
TURN RECORD TRACK
1
2
3
4
5
Game Turn
6
7
8
9
10

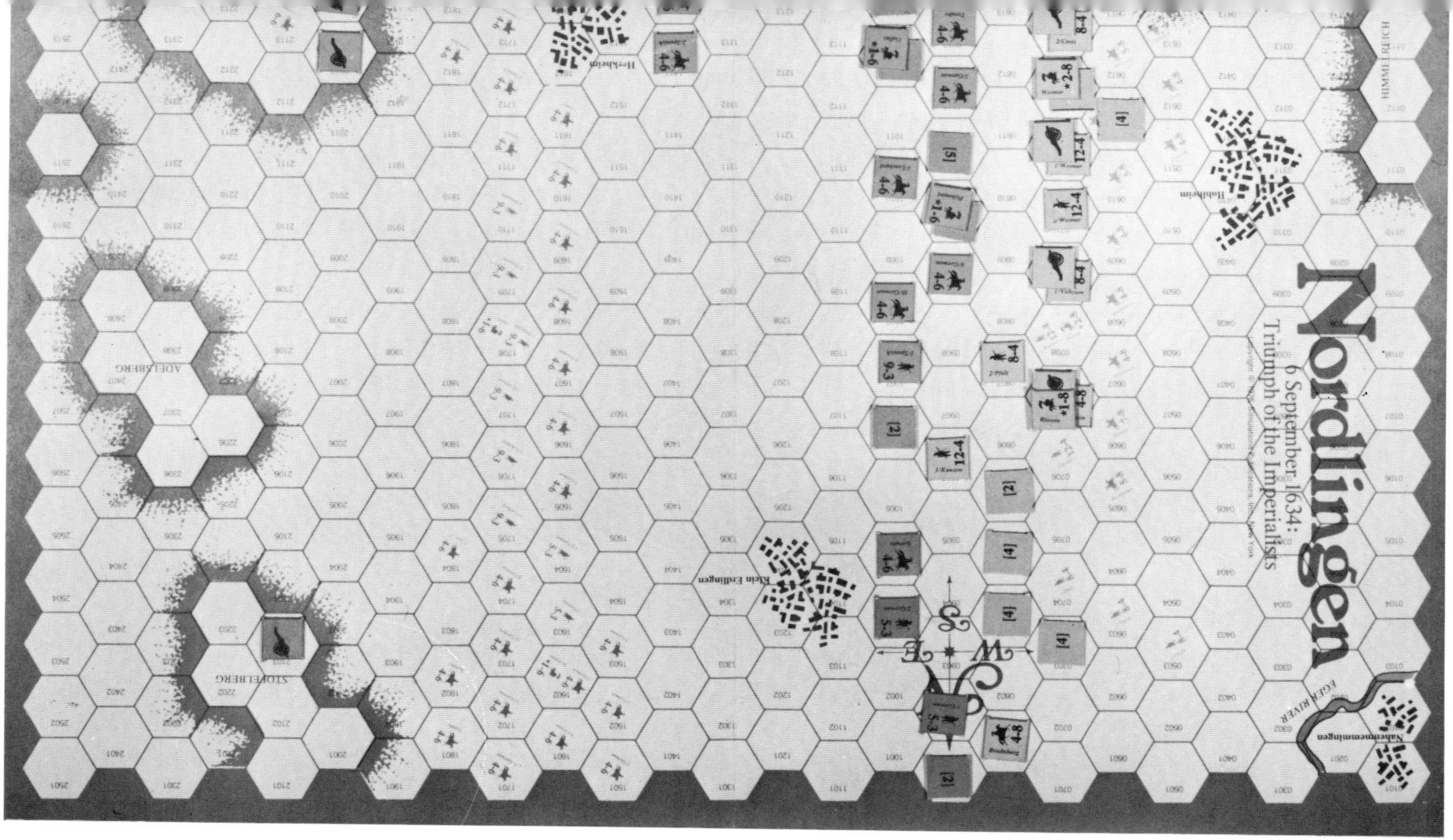

28 Turn 5. Position after Swedish turn.

nate in its die-rolls, and it might have gone differently if one or two key disruptions had not come off.

Where the Swedes scored decisively over the Imperialists was in pacing the flow of losses. They realized – and the Imperialists should have foreseen – that they had a single chance of victory on turn 5, and they grabbed it with both hands, passing the 100 points eliminated target with a unit to spare. The Imperialists pressed too hard on turn 4; had he made fewer attacks and got fewer disruption exchanges, the 100 points next turn would have been out of reach, while the Swedish 75 points might not have been.

The Imperialist reserves were well placed, and would have been used to good effect on the last fling on turn 5 if the position in general were not resignable. However, there was never much prospect that they could be usefully committed in the south once the Swedes sensibly retreated behind the Allbuch guns on the next turn.

The Swedish reinforcement of the southern position justified itself, though it is rather unusual. The weakening of the central front as a result was evident, but clearly counterbalanced by the victory on the Allbuch. All in all, the Swedish win (like most wins in *Nordlingen*) appeared more overwhelming after the fall of Imperialist morale than it really was – it *might* have gone the other way – but the margin of superiority was clear, if narrow, and the better Swedish planning earned the win.

Who were the players? Well, actually I played the game solitaire for this book. And I didn't look ahead sufficiently as the Imperialists to see what I would be able to do with the Swedes on turn 5. It's a wise man who can remember all his own good advice.

APPENDIX A: *Answers to Problems*

SOLUTION TO PROBLEM, CHAPTER 3

There are several small errors but two king-sized ones.
1) He hasn't read the victory conditions properly. The Mannerheim line is *not* worth ten points per hex. It's worth forty points if he captures the lot, otherwise nothing whatsoever. Rules mean what they say, not what the players think they meant to say.
2) The 6-4-2s in the centre can't reinforce the Petsamo attack, or anywhere else, except for the one nearest Petsamo; the rest are 4 turns' march from the rail line which they need to get them to the other sectors. Conversely, and still more important, nothing can reinforce the centre, because reinforcements start in the Soviet cities along the rail line. The handful of 6-4-2s has no chance of taking Oulu on its own. The Finnish 1-1-3s can perform their delaying tactics until the 4-4-2 arrives, plus the three 2-2-3s. With the help of the Finnish rail net, these can zip straight into the best defensive positions, and the 6-4-2s will be lucky to escape in one piece. Moreover, the 'gap' will be filled on the first Finnish turn by a 1-1-3 racing up the rail line, after which the prospects of the two 6-4-2s and 2-1-2s doing anything useful in 10 turns are minimal.

Other blunders: if it had been risky to leave the central area blank, then the Finns presumably wouldn't have done it, unless generalled by Smith himself. He is therefore being over-optimistic in thinking that some of the blank counters may be in the weak spots. But *anyway* it would scarcely matter, as reinforcements could loop around on the rail net on the first Finnish turn, from the other sectors. Then, how is Petsamo supposed to fall? Only two units can be put in front of it, as the third hex on one side is Norway (impassable), and the Russians can't get through the defending zone of control on the Soviet side to slip a third unit through (had Smith put a unit at the end of the peninsula past Petsamo, that would have been different). With his

29 Solution to problem Chapter 3.

inability to reinforce from the centre in a reasonable time span, he would have to attack the doubled 4-4-2 at 1–2 odds – which won't dislodge him. Fortunately for him, there is a 6-4-2 among the replacements, but he could certainly have used it elsewhere, and the odds will still be only 11–8, i.e. the dangerous 1–1.

Finally, the Ladoga line is not threatened at all, and the six 1-1-3s plus a blank have deterred the Russians from trying to outflank it. This means that the Finnish replacements can go to shore up either the Mannerheim line or Oulo, or even Petsamo! There is no way the Russians can win the game with these set-ups, and it is quite likely that they will fail to score at all.

Winter War is an enjoyable game, but it's important to realize from the start that the drive on Oulu has little chance against determined resistance, and the game will be won or lost in the fortified lines in the south. Petsamo should be winnable, but the Mannerheim Line must fall to get at least a draw. Smith should have massed his powerful units against the two fortified lines (if the Ladoga Line falls early, the Mannerheim position risks being outflanked; in any case, some Soviet commitment in this sector is necessary to protect the rail line for supply reasons), with a moderate force trying to outflank the Ladoga Line, and a number of small units making a diversionary drive on Oulu; the 2-1-3 should be employed here, as it has a little more mobility, which will make the defenders' task that much harder. The game can rarely be won for the Russians unless optional rules allowing Soviet paratroops are allowed – but these rules also open the possibility of the Finns calling a quick ceasefire under some circumstances, before the Soviet troops have got anywhere. With or without the options, the main effort *must* be in the south, to give a fighting chance.

SOLUTION TO PROBLEM, CHAPTER 4

Those damned paratroops! The weak spot is not the southern gap which I spent so much time checking, but the apparent stronghold of the Pripyat marshes. One armour and one infantry corps assault the 30th Soviet infantry (1-3) at 2–1 (the defence is tripled, but there are seven attack factors). *Meanwhile*, the paratroops go in against the 26th infantry (1–3) one hex behind, at 1–1. Perhaps the paratroops will die, but there is a good chance of taking the defenders with them, and then the road is open! Six more 4-6s pounce on the gap and pour out on the other side in the exploitation phase, racing up and down the rear of the line until almost the entire army is encircled by the German units and their ZOCs. A 4-6 on the west side of the line by the Baltic coast completes the net. The front is lined on the other side with 3-3s, set back where possible to avoid the isolated Russians, who are immobilized, from making any attacks before they have to surrender. There are a few attacks which the Russians can make, but they are all at unfavourable odds, and likely to fail. (See illustration 29.)

I don't like postal *Third Reich.*

SOLUTION TO PROBLEM, CHAPTER 5

England's letter is definitely false from start to finish. If he has been playing for four years, he is unlikely to have a strong preference for one alliance over another, and if he has, then why did he correspond with France first (this is an example of an intimidatory approach, meant to make you think that he has been diploming more than you, which backfires badly)? He asks you to attack France but makes no commitment to do so himself if you agree; France's letter suggests that England has tried the same tactic on him.

France's letter is a little too naive to be quite true, but much more reasonable. Note that he refers to England phoning him, whereas England spoke of a letter: this is a good sign, as it suggests they are not coordinating their letters to you in one of the very close alliances which sometimes appear.

Russia's piece sounds reasonable, though it doesn't give much away.

Italy's silence is a clear sign that he is not attacking France. This means Austria or Turkey are the target (unless Italy is just lazy), reducing the chances of an anti-Russian venture.

Austria's letter is probably genuine; he would not risk your telling Russia about his approach unless he was in earnest.

But is he right about Turkey? If Turkey were going north, *he* should be writing to you, since you can help both against Russia and later against Austria if he becomes too powerful.

You should therefore attack England and no one else; two-front wars are always unpromising, and it does not seem to be necessary against the others.

The letter to England should be written with great care, since any promise to him will probably be instantly relayed to France – he sounds that sort of player. Tell him you agree with his (unspecified) proposals, but want him to play a more active part. Add that you haven't heard from France. This will reassure France if the letter is sent on to him, as he will see that you were lying to England!

Tell France the absolute truth without hesitation or dissembling of any kind. You will help him as he suggests, and propose such-and-such a plan.

Tell Russia that there is an Austro-Turkish plot against him, enclosing Austria's letter as evidence. Accept his offer of cooperation in the North; if he believes in the plot, he will be happy to accept.

Write to Italy saying you have heard from Russia that he is going against Turkey. If Italy won't help you in the west, then you want him to attack Turkey, since the odds seem to be slightly on Turkey and Russia attacking Austria, and you want a deadlock down there while you forge ahead in the north and west.

Write to Austria expressing cautious interest, but saying you'll have to decide your policy at the last moment (which might under other circumstances be true). An Austrian attack on Russia suits you fine, but you don't want to spend your credibility by guaranteeing help which you know you will not provide. Warn him about possible dangers from Turkey or Italy.

Finally, drop Turkey an amiable note inviting long-term cooperation, and murmuring that Russia is believed to be going for him.

And write again to England the moment that the deadline for moves has past, explaining that it was a very difficult decision, but France seemed to make a better offer; if France has stabbed you after all, you are going to want to talk England round, so it is essential to keep diplomatic lines open.

SOLUTION TO PROBLEM, CHAPTER 6

Form a line of units one column east of the American regiments, starting with the hex (SS5) just south-west of the northern US unit in Monschau (2/9). Put 560/915 there; in the next hex down put 62/124 and both 340 regiments; in the next, 62/123 and both 26 regiments; in the next, 1SS Division; finally the two 18 units.

Commentary: The strength 4 infantry units in the attacks are of course interchangeable. There are 4 attacks: (1) 560/915 attacks 2/9 and 2/23 at 1–4. If this is engaged (1/6 chance) fine. If it is A elim, we are mildly sorry, but don't care much. A back results leave us indifferent. (2) 18 Division attacks 99/393 and 99/395 at the south end, at 1–1. We are delighted with either contact or (better still) engaged, as these results (1/2 chance) tie up the two enemy units between the attackers and the rough terrain; moreover, an engaged result will probably allow us to surround the enemy in open country next turn (they can counterattack, but it will be off a river, so halved – with luck the enemy will be eliminated if he tries it. A back allows the forces to escape directly south, while D back allows them to escape southwest. As our primary goal is to pin the enemy down, neither of these suits us, so there is no point in using more German units to get better odds – an even chance of contact or engaged is as good as we can get. It's a pity that

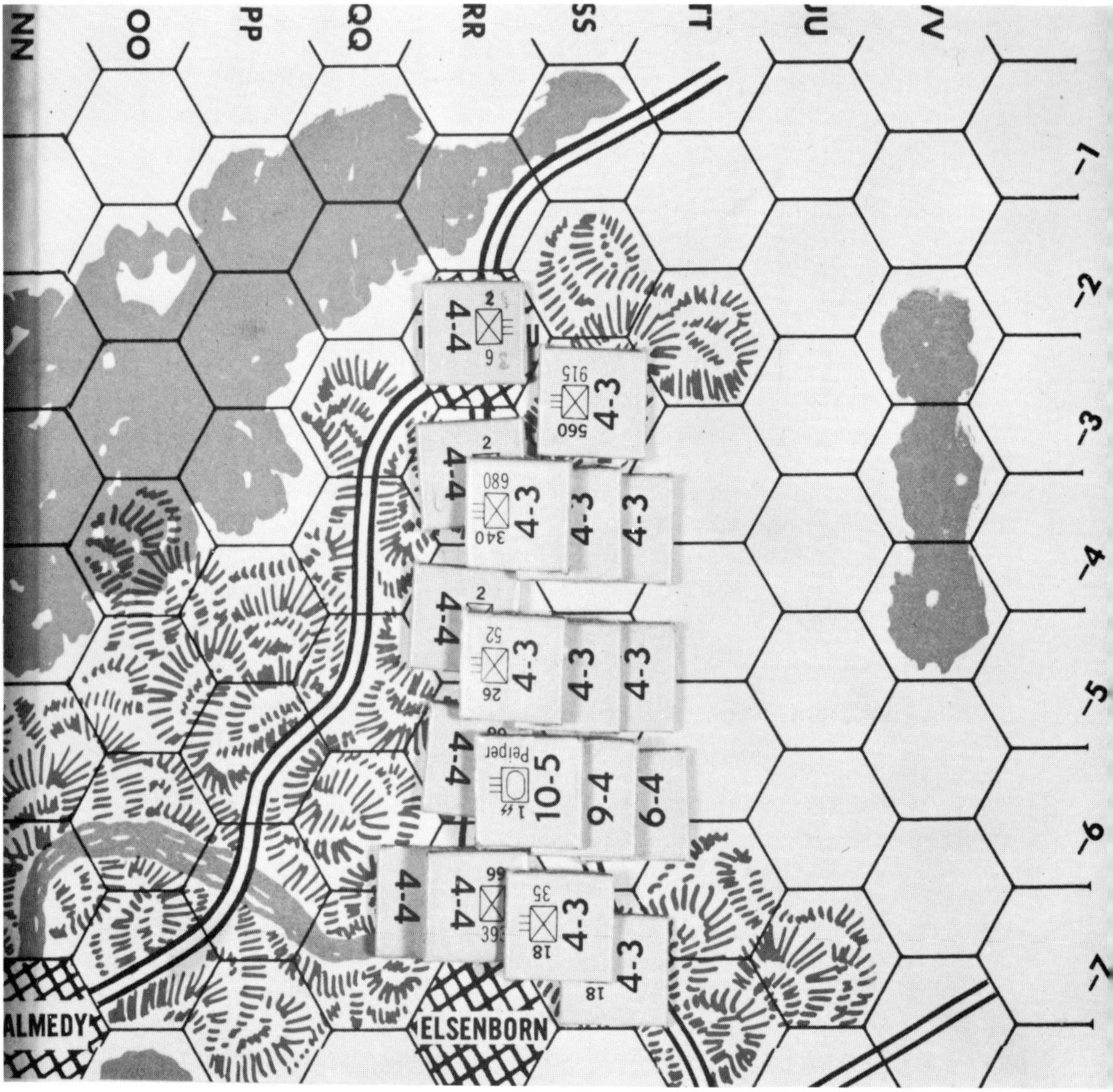

30 Solution to problem Chapter 6.

there is a 1/3 chance of A back, which will let the two units escape down the valuable road due south, but one can't have everything, and the forces required to attack at 2–1 (with the same chance of pinning, and A back only occurring one sixth of the time) are needed in the other battles. (3) 1SS Division and 26 Division attacks 99/361 at 4–1. There is a 5/6 chance of attacker advance here, which will seal off the 3 regiments to the north. (4) 62 and 340 Divisions attack 2/38 at 4–1. If attack 2 gave an attacker advance, then it is reasonable to reverse the order of 3 and 4 and divert 26 Division to the attack on 2/38, as this makes it certain that 2/38 will be forced back and 99/361 surrounded, with an even chance of being eliminated by an impossible retreat. Note that it doesn't otherwise pay to commit 26 Division against 2/38 (even though this increases the odds to 6–1 and only reduces the odds on 99/361 to 3–1), because the 4–1 on 99/361 is 1/6 more likely than the 3–1 to trap 3 units, and there is only a 1/36 chance (both die-rolls being 6s) that a 6–1 on 2/38 will be needed to trap the northernmost pair of units. This is a fine point and you needn't feel bothered if you did it the other way.

What you should have avoided, however, is a big attack at either end of the line, especially in the north, where it is almost useless to obtain an advance and positively helpful to the US to force the unit back down the south-west road. In the south, there is the excuse of blocking off the due south road, but as the main objective was pinning as many units as possible, the slackening of the grip on the units further north is too high a price.

SOLUTION TO PROBLEM, CHAPTER 7

My preference is for the following groups:

GERMANY
FORCE A, assault on wood: 3 Rifle, 5 SMG, 8 trucks
FORCE B, penetrating the line: 4 Panthers, 3 Pumas
FORCE C, defence of village: 5 150 mm, 3 SMG, 2 trucks.

For the attack on the wood, we can use the technique of dropping infantry in at one end and gradually pushing them through with close assaults. For this, attack factors and defence factors are both useful, so we have a mix of SMGs and riflemen. With luck, the 150 mms can give some support, but they may not be able to sight the area from their positions for the main job, Force C.

The penetration of the enemy line needs armour above all; as we are not trying to hold any captured points but simply break through, infantry would be relatively wasted here. To exploit the gaps which may appear in the line, we need fast tanks. Panthers combine a fair speed with considerable striking power. Pumas are much less powerful but very fast indeed, and can be used to overload a single defending unit at any critical defence point. Another Panther instead of two of the Pumas would be a reasonable alternative; exactly which is best depends on the Soviet choices.

The stacking limit means we can have only three infantry units in the village. Here, attack factors are crucial, because the enemy tactic will be to try and overload the defence, leaving one surviving unit to bring in murderous artillery fire, which will probably kill the defenders whatever their defence factors. To help against this tactic, the guns should be placed in cover in sight of the village. It is utterly pointless to leave anything (whether guns or more infantry or armour) in the open, as in the flat country specified they will get shot up immediately. The guns should be mostly H/M type to deal with the infantry likely to descend on the village. A good alternative to the 150 mm howitzers is the 120 mm mortars, which could be used together with CPs to fire, guided by the spotters, into other areas of the board. Against that, one would be paying for a range which is unnecessarily long in this situation, and the extra points of the 150 mms could be crucial: against a Soviet rifle unit the 150 mm has a 1–1 attack, while the 120 mm would give a tantalizing 15–16 = 1–2! The trucks may be able to move the guns (or some of them) into positions for other tasks if the battle for the village is over quickly.

I have rejected the massive Wespes (fine for pounding a single target, but too expensive when, as in all three tasks, the Germans are likely to have a number of targets at once), and the other armour (all rather too slow for these objectives, but too fast to just sit around as static artillery support).

SOVIET UNION
FORCE A, defence of wood: 5 SMGs.
FORCE B, holding the line: 8 76·2(A) guns, 3 T34cs.
FORCE C, assaulting the village: 5 rifle, 5 trucks, 5 120 mm(M).
3 CPs on hilltops surveying as much of the scene as possible.

The defence of the wood is easy: we need infantry to fight infantry, and no transport is needed as the forces will not be reassigned elsewhere. It is not worth having the 82 mm mortars to help, as they are so weak that the guns able to fire at any one time will have little effect, and cannot help infantry in close assault. As the wood's defenders will probably get first chance to attack, it is slightly more important for them to have a high attack factor. The converse of this odd paradox is also true. The units attacking the village will need to survive a turn in the open; for this, a good defence factor is vital, so rifle units are best there.

The line is best held by a series of anti-tank guns in each bit of cover, with the fastest Soviet tanks to reinforce any trouble spot.

Finally, the overloading operation is attempted with the riflemen and their transport, plus five powerful mortars to destroy the village if the spotters survive.

To find out what would happen between two forces, play it out as it would be likely to occur. In my case, it should go like this:

Battle for the wood: The eight German trucks bring in their force to an unoccupied part of the wood, and the infantry edges towards the enemy. As they get within two hexes, the defenders move out to meet them and get first attack. The total Soviet AF/DF is 40/60; the Germans' 39/54. With the advantage of first assault, the Russians have the edge unless the German 150 mm artillery can be diverted from defence of the village, which would put the result in the balance.

Penetration of the line: At least two of the 76·2 guns will have to be taken to provide cover for six of the seven German armoured units. At a doubled anti-tank strength of 24, this should lead to two Panthers being put out of action and later destroyed, plus further damage being done by the other defenders who can fire into cover guided by the defending 76·2. It seems unlikely if all the German armour can be taken out, however, though only a small force will break through.

Assault on the village: The five rifle units roar in to surround the village hex, with one of the trucks moving to the sixth neighbouring hex to provide an extra target. The defence is fairly hopeless. The best seems to be to hit five hexes with a 150 mm each, at 1–1, which gives an even chance of success. The lone truck can be destroyed by a close assault by one of the SMGs. The remaining SMGs will hope that there is only one (or at most two) of the attackers left to deal with, and close assault these. This is likely not to work, and on the next Soviet turn the 120 AF of the five 120 mms (plus whatever has survived intact) will almost certainly pulverize the village. It is therefore probably best for the Germans to switch their guns to support the attack on the wood, giving them a chance at coming out best in two of the three conflicts. The SMGs in the village, in this event, would be best advised to surrender in a hurry.

SOLUTION TO PROBLEM, CHAPTER 8

(A) The *Atlanta*, the *Enterprise* herself, and the *Vincennes* at the rear take on the T30 group, at odds of 30–12=2–1, giving 1, 2 or 3 hits.

Any other single escort fires at the T2 on the *Enterprise*'s port side, making this a 1–2, with no effect.

The remaining two ships fire on the D13, making this a 2–1, with 1, 2 or 3 hits.

The most probable result is a total of 4 hits on the *Enterprise*, leaving it afloat and the rest of the fleet unscathed.

(B) The secret of a good attack here is our old friend, overloading the defence. There are six ships, each of which can only fire on one group. Each carrier can be attacked from each side and from above, making six attacks and stretching the defence to breaking point. By committing six aircraft to each of these attacks, we will ensure that if any of these attacks are undefended, they will result in automatic victory at 6–1, while if one ship defends against each, we shall get three 2–1s against one carrier and two 2–1s and a 1–1 against the other (since five ships have flak factor 3, and the last one 6). This may well sink both carriers, and it would be amazing if neither sank: the average number of hits would be six and five respectively.

Not content with this, we can assign the remaining two torpedo-bombers and one dive-bomber to attacking the *Atlanta*, in three one-plane attacks from the three angles. If the American player fails to fire at one or preferably two of them, it is probable that the *Atlanta* will go down (if there is no flak it certainly will), which will be disastrous for the defence if one carrier remains afloat to receive a further attack later.

The best American strategy is probably to give up one carrier for lost, as well as the *Atlanta*, and concentrate the flak on the three groups of six planes attacking the *Yorktown*. The attacker can be restricted to a couple of hits this way, though the outlook without the two lost ships' flak cover remains very grim, unless the planes on the *Yorktown* can pull off a miracle against the enemy carriers, with the help of Midway planes.

It should be added that the attacker rarely has things quite so much his way!

APPENDIX B: *Useful Addresses*

MANUFACTURERS

Simulations Publications, Inc., 44 East 23rd Street, New York, NY 10010, USA (SPI)
The Avalon Hill Game Company, 4517 Harford Road, Baltimore, MD 21214, USA (AH)
Game Designers' Workshop, 203 North Street, Normal, IL 61761, USA (GDW)
Simulations Design Corporation, PO Box 19096, San Diego, CA 92119, USA (SDC, Conflict)
Philmar Ltd, 47–53 Dace Road, London E3 2NG
Wargames Research Group, 75 Ardingly Drive, Goring by Sea, Sussex, England (WRG)
Fact and Fantasy Games, PO Box 1472, Maryland Heights, MO 63043, USA
Jagdpanther Publications, PO Box 3565, Amarillo, TX 79106, USA
Metagaming Concepts, PO Box 15346, Austin, TX 78752, USA (MGC)
Lou Zocchi, 7604 Newton Drive, Biloxi, MS 39532, USA
Attack Wargaming Association, 314 Edgley Avenue, Glenside, PA 19038, USA (DCC–AWA)
Battleline Publications, PO Box 1064, Douglasville, GA 30134, USA
Excalibre Games, Box 29171, Minneapolis, Minn 55429, USA
Ironside Games, 133 Cherry Tree Road, Beaconsfield, Bucks, England
Tactical Studies Rules, 542 Sage Street, Lake Geneva, WIS 53147, USA
Jim Bumpas, 948 Loraine Ave, Los Altos, CA 94022, USA
Historical Perspectives, Box 343, Flushing Station, Flushing, NY 11367, USA
Maplay, 20 Kent Close, Orpington, Kent, England
Martial Enterprises, PO Box 1315, National City, CA 92050, USA

AGENTS

Unless their local shops stock the games, US readers will probably prefer to write to the companies direct. European readers may prefer to use local agents, since this saves both postage costs and delays. The following specialize in games from the companies named, though except where mentioned they also have games from other companies:

Avalon Hill UK, 646 High Road, North Finchley, London N12 0NL, England (Avalon Hill)
Games Centre, 16 Hanway Street, London W1A 2LS, England (GDW, SDC, Battleline, Third Millennia, Jagdpanther, Metagaming Concepts)
Simulations Publications UK, Freepost, Crown Passages, Hale, Altrincham, Cheshire, WA 15 6BR, England (SPI)
Charles Vasey, 5 Albion Terrace, Guisborough, Cleveland TS14 6HJ, England (Jagdpanther only)
Games Workshop, 97 Uxbridge Rd. London W12

As the situation in Europe outside the UK is developing rapidly, players in these countries should enquire the latest position, enclosing an addressed envelope and an IRC, from Walter Luc Haas, Postfach 7, CH-4024 Basel 24, Switzerland, who is the recognized authority on Continental wargaming, and sells many wargames himself.

CLUBS

There are three large groups which I know to provide a reliable postal wargames service; very likely there are others as well:
AHIKS, c/o Mike Truex, 304 White Road, Little Silver, NJ 07739, USA
Conflict Simulation Society, 39 Tweedstone Lane, Willingboro, NJ 08046, USA
National Games Club, 27 Elm Close, Amersham, Bucks, England

It should be emphasized that all three welcome members from any country.

MAGAZINES

Here again, there are numerous publications of varying quality and efficiency. I can recommend the following from personal inspection, but there is no doubt that others with which I am not familiar are equally good; *Outposts*, the magazine of the Conflict Simulation Society, and *The Kommandeur*, the equivalent from AHIKS, are frequently recommended, for instance, as is the magazine *Fire and Movement*.

PROFESSIONAL PUBLICATIONS

Games and Puzzles, 16 Hanway Street, London W1A 2LS, England. Monthly 50-page magazine on games and puzzles of every kind. I am Wargames Editor, and write several pages in each issue, particularly reviews of new games.

The General. The official Avalon Hill magazine, bimonthly. Useful for news of the company's activities, but above all read for its excellent Series Replays of AH games blow-by-blow, and in-depth analyses of various games each month in the AH line. *See* AH under Manufacturers.

Battlefield, the Jagdpanther magazine in which the games first appear. See Jagdpanther under Manufacturers.

Campaign, PO Box 896, Fallbrook, CA 92028, USA. Independent magazine writing about the products of all the producers. Bimonthly.

Moves, the SPI equivalent of *The General*, with similar advantages. *See* SPI under Manufacturers.

Strategy and Tactics, another SPI bimonthly, with one of their games (later to be sold separately) free in each issue, with historical and design notes, plus articles on other games, and the results of polls of readers. Distributed in Britain with *Phoenix*, the journal of SPI-UK, which is also available separately.

AMATEUR PUBLICATIONS

Battleground, editor Marcus Watney, 22 Alexandra Road, Reading, Berks, England. Small but packed monthly, with the advantage of offset printing allowing frequent illustrated Series Replays. Organ of the NGC wargames section; I am biased in its favour as I was its first editor. The editor is a professional journalist, which shows in the quality of text and layout.

Europa, editor Walter Luc Haas, Postfach 7, CH-4024 Basel 24, Switzerland. Enormous, appearing at irregular intervals, with articles on every conceivable aspect of wargaming from some of the leading figures in the hobby. Essential reading for any serious player.

Perfidious Albion, co-editors Charles Vasey and Geoff Barnard, 5 Albion Terrace, Guisborough, Cleveland TS14 6HJ, England. Monthly specializing in reviews, variants and a fine line in analyses of historical accuracy.

Owl and Weasel. Magazine of Games Workshop (see list of agents) with special emphasis on fantasy games.

Walter Luc Haas also publishes a German-language companion magazine to *Europa*, with a similarly encompassing range of topics.

APPENDIX C
Unit values in Panzerblitz *and* Panzerleader *free-choice selection*

Some years ago, Tom Oleson wrote an article entitled *Situation 13* for the Avalon Hill *General* magazine, which suggested a method for choosing units in *Panzerblitz* according to a points system rather than using one of the twelve scenarios in the game. Each side could choose up to (say) 1000 points' worth, and these forces would fight it out.

My experience in the National Games Club is that this transforms the game to such advantage that no *Panzerblitz* player ever plays a regular situation again. Used with club hidden-movement rules, the effect is particularly thrilling, since you have no idea what kind and size of force you are going to run into – 20 powerful tanks, or 60 weak units with trucks and towed guns, or some mixed force. However, even the open-movement game is decisively improved, with each game different from all the others. In view of the succcess of these rules, Geoff Barnard, wargames organizer of the club, devised open- and hidden-movement versions for the companion game, *Panzerleader*. Since the games are widely played, but the free-unit-choice system is not as well known as it should be, there follows a brief description of each of the open-movement versions, which should enable the reader to play them if he has the games. I am indebted to Avalon Hill and Geoff Barnard for permission to describe the systems; there are some amendments to the *Panzerblitz* one which the club has found useful, but basically the design remains the idea of Tom Oleson.

Each player chooses his force in secret from the units available on his side. No forts are allowed in *Panzerblitz*, and no air units, flail tanks, or bridges in *Panzerleader*, the specialized values of these being hard to quantify. The Turreted Vehicles experimental rule in *Panzerleader* should be omitted, as it would change point values; in *Panzerblitz* the use of all optional but no experimental rules is recommended.

After the force choice, the placing of the geomorphic boards is determined at random, and both players set up simultaneously on their respective home boards. Mines and blocks, however, may also be placed on the nearer half of the middle board. Hexes half on each of two boards may be counted either way as convenient at the time, but you cannot change your mind thereafter. The player with fewest units moves first. The winner is the side with most units on the central and enemy boards by game end, with units moving off the enemy home edge counting double (but such units may not return). There are twelve turns.

You cannot have more units than there are in the box, or more than half the blocks or mines (a practical convenience, and necessary to stop a total stonewall strategy). Mines and blocks do not count either for deciding who starts or for determining the winner.

The reason for using unit count rather than point count for victory is that the smaller force has two advantages: it gets to the centre first, and it must be more powerful, since both sides had 1000 points available. Generally speaking, moderately large forces seem to work best, with mixed unit types. See the Combined Arms chapter for further discussion.

Unit values:

GERMANY		**SOVIET (PANZERBLITZ) and ALLIED (PANZERLEADER)**			
UNIT	VALUE	UNIT	VALUE	UNIT	VALUE
Guns		*Guns*		*Guns*	
50 mm	9·5	12·7 mm	8	37 mm	9 (US/GB)
75 mm A	11·5	45 mm	8	76 mm	13 (US/GB)
88 mm	21	57 mm	9·5	90 mm	18·5 (US/GB)
20 mm	8	76·2 mm A	10·5	17 pdr	14 (GB)
20 mm (Quad)	13	76·2 mm H	8·5	40 mm	12 (US/GB)
75 mm H	9	122 mm	31	25 pdr	54·5 (GB)
150 mm	18	82 mm MOT	11	105 mm	58 (US)
81 mm	11·5	82 mm	11·5	155 mm	80 (US/GB)
120 mm	19·5	120 mm	24	8 in	102 (GB)
37 mm	10			76 mm M	9·5 (GB)
75 mm IG	9			81 mm	11·5 (US)
75 mm How	36			107 mm	15·5 (US/GB)
105 mm	58				
150 mm H	80				
170 mm H	92				
Nebekwerfer	69				
Infantry					
Engineer	18 (-*leader* 16)	Engineer	16	Engineer US; 9 GB: 16	
Security	9 (-*leader* 7·5)	Reconnaissance	9	Scout	5·5 (US/GB)

UNIT	VALUE	UNIT	VALUE	UNIT	VALUE
Infantry					
Rifle	13 (*-leader* 11)	Rifle	23	MG	7 (US/GB)
SMG	14 (*-leader* 12·5)	SMG	22	Rifle	9 (US/GB)
CP	5	CP	5	Amd inf	14 (US/GB)
		Guard	26		
Transport, etc.					
Wagon	4	Wagon	4	M3 Scout car	16 (US/GB)
Truck	7	Truck	7	Truck	8 (US/GB)
Halftrack	14	Halftrack	12	Halftrack	13 (US/GB)
		Cavalry	20	Bren carrier	12 (GB)
Armoured cars					
Puma	28			M20	21 (US/GB)
SdKfz 234/4	38			M8	26 (US)
SdKfz 234/1	23			Daimler	27 (GB)
Self-propelled artillery					
Maultier	70			Sexton	67·5 (GB)
Wespe	69			M7	71 (US)
Hummel	86			Recon HQ	24 (GB)
Assault guns					
Gw 38	27	SU 152	68	M16	19 (US)
Wirbelwind	33			M4/105	36 (US/GB)
Stu H 42	40			Churchill	57 (GB)
Tank-destroyers					
Marder III	35	SU 76	35	Achilles	41 (US/GB)
StuG III	40	SU 85	45	M10	39 (US)
Hetzer	38	SU 100	47	M18	40 (US)
JgdPz IV	45	JSU 122	49	M36	42 (US)
Nashorn	54				
JgdPz V	56				
Jgd Pz VI	57				
Tanks					
Lynx	22	KV 85	46	M5	26 (US/GB)
PzKpfw III	30	T 34c	38	M24	37 (US)
PzKpfw IV(W)	34	T 34/85	44	Cromwell	32 (GB)
PzKpfw IV(SS)	38	JS II	46	Sherman	34 (GB)
PzKpfw V(W)	46	JS III	52	M4/75	36 (US)
PzKpfw V(SS)	50			M4/76	43 (US)
Tiger I	47				
Tiger	54				

Mines are 10 points each; blocks 2 each.

The *Panzerblitz* PzKpfw IV–V types are the SS ones (the latter is better known as the Panther). Most points values stem from the sum of the four combat factors, with modifications for generally useless factors such as rifle range (most rifle units in the games fight at close quarters or not at all, especially in *Panzerblitz*).

For information on the hidden-movement versions, contact the National Games Club (wargames section), enclosing four international reply coupons, or (UK readers) 40p. The address is c/o Geoff Barnard, 4 Albion Terrace, Guisborough, Cleveland TS14 6HJ, England.

NOTE

Introduction to Strike Force One

For the benefit of readers unfamiliar with board wargames, SPI's *Strike Force One* is enclosed at the back of the book. This game, kindly supplied free by the manufacturers for this book, is available in further copies for the postage cost from the company.

It is suggested that the game be played in conjunction with Chapter 2, in the introductory part of the book: the elementary concepts discussed there are shown in practice in *Strike Force One*. It should be emphasized that the game, with its ten units and tiny board, is designed to illustrate the basic system used by the overwhelming majority of wargames, and is not intended to be a fully fledged game on the scale of those described in Part IV. But I hope it will give new players a pleasant introduction to the fundamental ideas of wargames.